Hot Science is a series exploring the ~~~~~~~
and technology. With topics from big data to rewilding,
dark matter to gene editing, these are books for popular
science readers who like to go that little bit deeper ...

AVAILABLE NOW AND COMING SOON:

Destination Mars:
The Story of Our Quest to Conquer the Red Planet

Big Data:
How the Information Revolution
is Transforming Our Lives

Gravitational Waves:
How Einstein's Spacetime Ripples Reveal the Secrets
of the Universe

The Graphene Revolution:
The Weird Science of the Ultrathin

CERN and the Higgs Boson:
The Global Quest for the Building Blocks of Reality

Cosmic Impact:
Understanding the Threat to Earth from Asteroids
and Comets

Artificial Intelligence:
Modern Magic or Dangerous Future?

Hot Science series editor: Brian Clegg

ASTROBIOLOGY

The Search for
Life Elsewhere
in the Universe

ANDREW MAY

ICON

Published in the UK in 2019
by Icon Books Ltd, Omnibus Business Centre,
39–41 North Road, London N7 9DP
email: info@iconbooks.com
www.iconbooks.com

Sold in the UK, Europe and Asia
by Faber & Faber Ltd, Bloomsbury House,
74–77 Great Russell Street,
London WC1B 3DA or their agents

Distributed in the UK, Europe and Asia
by Grantham Book Services,
Trent Road, Grantham NG31 7XQ

Distributed in the USA
by Publishers Group West,
1700 Fourth Street, Berkeley, CA 94710

Distributed in Australia and New Zealand
by Allen & Unwin Pty Ltd,
PO Box 8500, 83 Alexander Street,
Crows Nest, NSW 2065

Distributed in South Africa
by Jonathan Ball, Office B4, The District,
41 Sir Lowry Road, Woodstock 7925

Distributed in India by Penguin Books India,
7th Floor, Infinity Tower – C, DLF Cyber City,
Gurgaon 122002, Haryana

Distributed in Canada by Publishers Group Canada,
76 Stafford Street, Unit 300
Toronto, Ontario M6J 2S1

ISBN: 978-178578-342-5

Text copyright © 2019 Icon Books Ltd

The author has asserted his moral rights.

No part of this book may be reproduced in any form, or by any
means, without prior permission in writing from the publisher.

Typeset in Iowan by Marie Doherty

Printed and bound in the UK by CMP Books

ABOUT THE AUTHOR

Andrew May is a freelance writer and science consultant. He has written on subjects as diverse as the physical sciences, military technology, British history and the paranormal. His recent books include pocket-sized biographies of Newton and Einstein, an eye-opening study of the relationship between pseudoscience and science fiction, and *Destination Mars* and *Cosmic Impact* in the Hot Science series. He lives in Somerset.

CONTENTS

LIFE BEYOND EARTH

1

In July 2018, the UK tabloid newspaper the *Daily Express* carried a story dramatically headlined 'Aliens on Europa: NASA Hunts for Life Just 1 cm under Surface of Jupiter's Moon'. To reinforce the message, it was accompanied by an eye-catching composite image. On one side of the graphic was a photograph of Europa: an enigmatic-looking world, nothing like our own Moon, with a smooth surface of solid ice, criss-crossed by dark cracks. On the other side of the image, an artist's impression of a typical 'alien': grey-skinned but otherwise distinctly humanoid in appearance, with a high-domed, hairless forehead, large eyes and delicate features.

Confusingly, however, the article's strapline read as follows:

> Scientists hoping to find alien life on Jupiter's moon Europa may not have far to search after a study revealed microbes could be surviving just one centimetre beneath the surface.

So what's going on? Is NASA going to Europa to hunt for big-brained, humanoid aliens, or for tiny little microbes? Delving further into the fine print, it turns out the answer is neither. The *Express* article was prompted not by a space mission that's about to take off, but by a clever piece of scientific deduction. It is widely accepted that, if life exists on Europa – and yes, we're most likely talking about microscopic organisms here – it's to be found in the ocean of liquid water believed to exist several kilometres below the icy surface. The new development the *Express* picked up on is the suggestion that chemical traces of life – for example proteins or complex DNA-like molecules – might be found near to the surface of the ice, making them much easier for a space probe to detect. This idea originated in a scientific study that had just been published in the journal *Nature Astronomy*, which came to the following conclusion:

> These results indicate that future missions to Europa's surface do not need to excavate material to great depths to investigate the composition of endogenic material and search for potential biosignatures.*

This is real science, and in principle it's no bad thing that it found its way into a widely read tabloid like the *Daily Express*. But the way the newspaper chose to report it – and the way the popular media treats stories of this kind in general – is likely to leave readers more confused than enlightened. Are

---

* 'Endogenic' is just a fancy way of saying 'formed underground'. As for 'biosignatures' – hold that thought for a few pages and all will be explained.

they saying that NASA believes there are humanoid aliens on Europa? Why go all the way to Europa when the same newspaper also frequently reports anecdotal sightings of humanoid aliens here on Earth? Is NASA on the point of sending a space probe to look for life on Europa, or is that just an idea for the future? Why do scientists keep going on about extraterrestrial microorganisms, when everyone knows that aliens are pretty much like us except for their big eyes and high foreheads?

All these things – and many more – will be clarified in the course of this book. Astrobiology is a wide-ranging subject, dealing with the possibility of life beyond Earth from every conceivable angle. To start with, however, let's kick off with a much simpler question.

Is There Life on Earth?

From our perspective on the surface of the planet, it's obvious there's life on Earth. From out in space, too, it's not that difficult to detect. The night side of the planet is lit up by city lights, there are thousands of small artificial satellites in orbit, the radio spectrum is buzzing with structured signals that have no natural explanation, and the atmosphere is laced with industrial pollutants.

But all those things have existed for just a century or so: a tiny fraction of the Earth's lifetime, which is about 4.5 billion years. Nevertheless, life – at a less obvious level – has existed for a significant fraction of that time, perhaps as much as 4 billion years. Until just under a billion years ago, all of that life (and the vast majority of it even today) took the form of tiny single-celled organisms – the 'microbes' that scientists are so fond of talking about. The following table shows how, over the course of time, increasingly complex forms gradually evolved and were added to the mix of life on Earth.

Milestones in the evolution of life on Earth (all dates are approximate)

Time before present (in millions of years)	Evolution of life on Earth
4,500	Formation of the Earth
4,000	First single-celled life forms
1,700	Microscopic multicellular organisms
600	Small marine animals; seaweed
450	Fish; land plants; insects
350	First land vertebrates (amphibians); trees
180	Jurassic dinosaurs; flowering plants
50	First lemur-like primates
3.5	First tool-making hominids
0.3	*Homo sapiens*

This means the question of life on Earth is a matter of definition. To a scientist, 'life' includes any kind of living thing – even if it can only be seen through a high-power microscope. By that definition, Earth has been home to life for almost 90 per cent of its history. On the other hand, people brought up on a diet of sci-fi movies and tabloid stories about UFO encounters are more likely to equate 'life' with a technologically savvy civilisation – in which case that 90 per cent figure drops all the way down to 0.000002 per cent.

If we're going to look for life on other Earth-like planets, what are the relative chances of finding it by those two definitions? We can make a rough estimate by picking random snapshots of the Earth at different points in its 4.5 billion-year history. On that basis, the chance of finding life – by the ufologist's or sci-fi fan's definition – is so tiny as to be virtually zero. By the scientist's definition, on the other hand, the chances are pretty good.

So let's look a bit more closely at that 'scientist's definition of life'. The nature of life turns out to be surprisingly difficult to pin down, and precise definitions tend to vary between specialists working in different branches of science. As far as astrobiology is concerned, a good starting point is the working definition devised by NASA in the 1990s:

Life is a self-sustaining chemical system capable of Darwinian evolution.

That's refreshingly concise, but it packs a lot into a small number of words. The first part, 'self-sustaining chemical system' is clear enough. But the latter part, 'capable of Darwinian evolution', hides a lot of detail. It doesn't just mean that our self-sustaining chemical system has to be able to evolve, or change its form over time. First, there's an implicit assumption that the change occurs over successive generations, each of which is born, grows and dies. Then there's that word 'Darwinian' – after Charles Darwin, the Victorian naturalist who did far more than suggest that living species evolve. He argued that they do this for a reason – to adapt to the changing circumstances of their environment – and that they do so by means of natural selection, or 'survival of the fittest'.

The beauty of this definition is that it encompasses everything from the single-celled organisms that emerged on Earth 4 billion years ago – and may possibly be hiding under Europa's ice sheets – via semi-civilised primates like ourselves, all the way up to super-advanced lifeforms we can hardly even imagine. Astrobiology – the subject of this book – deals with the possibility of life beyond Earth wherever it falls in that spectrum. As the 'astro' prefix implies, it's essentially a sub-branch of astronomy, using the same sort of telescopes, space probes and theoretical techniques that astronomers apply to any other facet of outer space.

Earlier in this chapter (page 2) we saw a quote from a scientific paper featuring a lot of multisyllabic words. One of them, 'biosignatures', will turn out to be one of the most important words in this book. A moment ago, we saw how all the obvious ways an outside observer might detect life on Earth – artificial lights, satellites, radio signals, etc. – relate to our own civilisation. But there are other, subtler, ways of detecting more primitive lifeforms – and these are collectively known as biosignatures. Most importantly, living organisms produce, as waste products, tell-tale chemicals that would be very difficult to account for in terms of non-living processes. These chemical 'signatures' are potentially detectable to astro-biologists through telescopes or spacecraft-based sensors.

At the upper end of the spectrum of life, biosignatures are joined by 'technosignatures': detectable indications of a technologically advanced civilisation. As we'll see later in this book, there are numerous possibilities here, but per-haps the most obvious – and the easiest for us to recognise as artificial – would be some kind of deliberate interstellar communication. In a historical context, the first practical efforts in astrobiology were aimed at detecting such communications, under the name of SETI – for 'Search for Extraterrestrial Intelligence'. SETI is still going strong, although confusingly it uses the word 'intelligence' in a different way from people working in other branches of science.

To a biologist or psychologist, intelligence is the capacity for understanding and logical reasoning. By this definition, human beings were every bit as intelligent thousands of years ago as they are today. Yet from a remote-sensing point of view, they didn't produce any detectable signatures that were noticeably different from far more primitive animals. So, as insulting as it is to, say, Alexander the Great or Lao Tzu or Akhenaten, they simply weren't 'intelligent' by the standards of SETI researchers. They only produced biosignatures, not technosignatures.

Since I've started to quibble about other people's choice of words, here's another thing. Although SETI is a sub-branch of astrobiology, who's to say that a SETI signal – if and when it's detected – necessarily has a biological origin? It might be the work of an advanced AI – artificial intelligence – which has outlived its organic creators. Whether such an AI constitutes 'life' is a question for the philosophers – but we can say right away that it doesn't conform to NASA's definition. It's not a 'chemical system', and it's almost certainly the result of intelligently driven evolution rather than Darwinian natural selection.*

---

* For that matter, even advanced biological species are likely to control their own evolution, via genetic engineering, rather than leaving it to Darwin.

We can think of biosignatures and technosignatures as overlapping sets. The first is looking for biological life of any kind (technological or not), the second for technological civilisation of any kind (biological or not). Judging from the situation on Earth over the last several billion years, we might conclude that the first has a good chance of success, while the second is like searching for a very small needle in a very large haystack.

Fortunately, the prospect for technosignatures may not be as bleak as that. We're forgetting that Earth has – hopefully – several billion years of existence ahead of it. Who knows what might happen in that time: a technological society that's as far ahead of us as we are from the stone age, or a post-human world ruled by computers, or in which people have 'uploaded' themselves into digital form and can whizz around the galaxy at the speed of light?

As unimaginably ancient as 4.5 billion years sounds to us, the Earth is really quite young in a galactic context. The oldest Earth-like planets are likely to be around twice that age, while the average age is probably around 6 billion years. With a head start like that, the galaxy could be teeming with super-advanced aliens.

Fermi's Paradox

Enrico Fermi was one of the most important scientists of the 20th century. He won the Nobel Prize in 1938 for his work on nuclear physics, and during the Second World War he was part of the team at the Los Alamos laboratory in New

Mexico where America's first atomic bombs were built. After the war, Fermi took up a professorship at the University of Chicago, but continued to make regular trips back to Los Alamos, where he acted as a consultant during the development of the ultimate Cold War weapon, the hydrogen bomb.

On one such visit in the summer of 1950, Fermi got into a lunchtime discussion with colleagues that had nothing to do with nuclear physics. This was just three years after the media had coined the term 'flying saucer' to describe alleged sightings of alien spacecraft, and the papers were still buzzing with stories about them. It seems this was the topic the Los Alamos physicists were discussing that day. They were particularly amused by a cartoon in the *New Yorker* magazine, which attributed a phenomenon described in a completely separate news story – mysteriously disappearing trash cans – to alien visitors.

As scientists, they were unconvinced by all the supposed evidence for flying saucers, because they could see there were always other, more likely explanations. At the same time, they were aware of the vast scale of the galaxy – both in terms of its enormous age and the sheer number of stars – and realised that, in fact, there ought to be extraterrestrials everywhere. Fermi summed up the problem with a simple question: 'Where is everybody?'

It's gone down in history as Fermi's Paradox: the idea that space should be teeming with aliens, and yet we see no evidence of them. But is that really a paradox? To some people, it's far from obvious that 'space should be teeming with aliens', while UFO believers would scoff at any suggestion that 'we see no evidence of them'. More subtly, there's an unspoken assumption: that if aliens do exist, we ought to see evidence of them – which is just as contentious. The plain fact is that one of these three statements *must* be false. There's a flaw either in the logic that says extraterrestrial intelligence must be widespread, or in the assumption that we ought to be able to detect it, or in the assertion that there's no evidence for it. We'll need to learn a lot more about the subject before we can address the first two points – we'll come back to them in Chapter 4 – but we can deal with that last point right away.

Is it really true that there isn't a shred of observational evidence suggesting the existence of aliens? There's certainly no shortage of people prepared to dispute that. The very term 'unidentified flying object' (UFO) has come to be synonymous with 'extraterrestrial spacecraft'. Now, it's a well-documented fact that witnesses – including professional pilots and astronauts – occasionally see flying objects they can't identify. So no one can seriously deny that unidentified flying objects exist. Isn't that the same as saying that no one can deny the existence of alien spacecraft? To anyone who makes the effort to parse each of those words separately (unidentified – flying – object), no, it's not the same thing at all – but to others the two statements are completely equivalent.

There are plenty of other 'mysteries of the unexplained' that can be accounted for by aliens, too. Something crashed at Roswell, New Mexico in June 1947, and the military initially described it as a flying saucer – even if they later retracted the claim – so isn't that proof enough? Who taught the ancient Egyptians how to build giant pyramids, if it wasn't visitors from another planet? Even if aliens weren't responsible for those disappearing trash cans in New York, surely they must have engineered the utterly inexplicable disappearance of that Malaysian airliner in 2014?

Questions like these are highly divisive between the scientific and UFO communities. To a scientist, they're questions that can all be answered in more mundane, and far more likely, ways. To believers, the existence of alternative explanations is irrelevant. If it *might be* aliens, then it *is* aliens. The TV series *The X-Files* hit the nail on the head with the phrase 'I want to believe'. If a person wants to believe a theory badly enough, they end up seeing evidence for it everywhere.*

Having said that, it would be arrogant to suggest that just because an idea is 'unscientific' it's necessarily wrong. That's not something I would ever say about someone's spiritual beliefs, because they're almost always framed in a way that's impossible to prove or disprove by scientific methods. The situation in the UFO community is very similar. Either by accident or design, they've built up an edifice of belief that can never be proved wrong. Their aliens are powerful enough to manipulate anything that might be used in evidence against them: our perceptions, memories, computer data – possibly even reality itself. They may even be deliberately mischievous, ensuring that believers see enough evidence to continue to believe, while sceptics and unbelievers see no evidence at all. As outrageous as that sounds to a

---

* I wrote about this in my book *Pseudoscience and Science Fiction* – as the book's title suggests, it's a worldview that draws much more heavily on sci-fi and popular culture than on an understanding of science.

rational mind, it's not impossible – but it's an idea that falls in the realm of philosophy, not science.

By its nature, science can only deal with objective, repeatable evidence. Since that's what this book is about, we'll have to wave goodbye to UFOs and their mischief-making occupants at this point. Fortunately, the question of whether or not they exist doesn't alter the validity of what we're going to say about the search for biosignatures and technosignatures.

The Way Forward

To have any hope of making progress, astrobiology has to focus on practicalities – constructing instruments, making observations, interpreting data – rather than just sitting and thinking about the subject, which is what philosophers and sci-fi writers do. There's nothing wrong with speculation, but it only turns into real science if that speculation comes up with consequences that can be tested by observation. Going back to the Europa example, for instance, the notion of microorganisms living in a subsurface ocean there is speculation, but the idea that their chemical signature might be detectable to a spacecraft sitting on the surface is real science.

Then again, a book of this type can't overlook the speculative side altogether, because the question of life beyond Earth is a multidisciplinary one that attracts the interest of philosophers, theologians, Hollywood screenwriters and tabloid journalists as well as scientists. So we'll take a look at some of those broader aspects in the next chapter, and then visit them again at the end of book. The science can be found in Chapters 3 and 4, dealing with various techno-signatures we might expect to see from advanced alien civilisations, and Chapters 5 and 6, on simpler lifeforms and their biosignatures.*

---

* That might sound back-to-front, but the science of biosignatures is actually the newer, and more complex, of the two.

THINKING ABOUT ALIENS

2

The concept of extraterrestrial life is inextricably tied to our modern-day picture of the universe, in which the stars are distant suns, many of them – like our own Sun – with a retinue of planets orbiting around them. But people haven't always seen things that way. Before the invention of the telescope, all the stars in the night sky were nothing more than tiny points of light, seemingly attached to the inside surface of a sphere that revolves around us. The only object that might conceivably have been another world like ours was the Moon. To the philosophers of ancient times, that made it the most likely place to find 'extraterrestrial life'. The Greek historian Plutarch, writing almost 2,000 years ago, attributed the idea to followers of Pythagoras, who had lived several centuries earlier still:

> The Pythagoreans affirm that the Moon appeareth terrestrial, for that she is inhabited round about, like as the Earth wherein we are, and peopled as it were with the greatest

> living creatures, and the fairest plants ... There is nothing
> that doth prove and show directly, this habitation of men
> in the Moon to be impossible.

When the first telescopes were directed at the Moon in the 17th century, they revealed an intriguing landscape of mountains, valleys and craters that had only been hinted at by naked-eye observations. This prompted a young English clergyman, John Wilkins, to write a book called *The Discovery of a World in the Moon*. Published in 1638, the book drew numerous parallels between the Earth and the Moon, leading Wilkins to conclude that:

> 'Tis probable there may be inhabitants in this other world,
> but of what kind they are is uncertain.

That simply echoes the ancient Greeks, but Wilkins adds a remarkable insight of his own: 'As their world is our moon, so our world is their moon.' That was a hugely important step in our understanding of Earth's place in the cosmos: that there is nothing unique or special about it. Wilkins made the same point even more clearly in his next book, published two years later. Its attention-grabbing title was *A Discourse Concerning a New Planet*.

Anyone familiar with the history of astronomy will know that no 'new planet', beyond those visible to the naked eye, was discovered prior to Uranus in 1781, a century after Wilkins's time. But that's not what he's talking about. The point he's making is clarified in the book's subtitle: 'Tending to prove that 'tis probable our Earth is one of the planets'.

A

DISCOURSE

Concerning a

New Planet,

Tending to prove
That 'tis probable our E A R T H
is one of the P L A N E T S.

**The strikingly worded title page of a book
by John Wilkins, written in 1640.**

(Public domain image)

Today, it's blindingly obvious that the Earth is a planet.
It's the very archetype of a planet, the one against which
all the others are measured and contrasted. But things
were different in Wilkins's day. To most of his readers, the
word 'planet' would have meant one of five points of light –
Mercury, Venus, Mars, Jupiter or Saturn – in the night sky.
The word comes from *aster planetes*, the Greek for 'wandering
star'. To the unaided eye, there is almost nothing to distinguish these planets from other bright stars, except for the
fact that they slowly drift, from one night to the next, against
the background of seemingly fixed stars.

The idea that Wilkins was promoting – that the Earth is
just another of these planets, and that they all revolve around
the Sun – had been developed a century earlier by Nicolaus
Copernicus, an administrator at Frombork cathedral in
Poland. It was one of the great scientific revolutions, displac-
ing the Earth from its privileged position at the centre of the

universe.* In more general terms, the 'Copernican principle' – that there is nothing special about our cosmic location – has become one of the fundamental tenets of astronomy. Yet Copernicus himself only took the first step in that direction. Like most of the ancient and medieval scholars who preceded him, he continued to picture the 'fixed' stars as being tiny objects attached to the interior of a sphere surrounding the Solar System, and nothing like our own Sun in terms of size and luminosity. It was left for others to dispel that illusion.

The Plurality of Worlds

Thomas Digges was born in 1546, three years after Copernicus published his revolutionary theory. The son of an astronomer, Digges encountered Copernicus' work at an early age. By the time he was 30 he had produced the first account of it in the English language: *A Perfect Description of the Celestial Orbs*. But Digges went even further than Copernicus. He conjectured that the stars, rather than being fixed to a cosily Sun-centred sphere – 'as in a nutshell', as he put it – are spread out through an infinite universe.

Given this impressive insight, it's sad that history has all but forgotten Digges. That can't be said of a young playwright who once lodged in the same house as Digges in London's Bishopsgate – one William Shakespeare. The two must have

---

* Actually, Copernicus seems to have been an unlikely revolutionary, motivated more by a desire to predict celestial movements to a few more decimal places than to sweep away everything humans believed about their place in the scheme of things.

had some fascinating conversations, and Shakespeare drew on them many years later when he came to write one of his most famous works, *Hamlet*. At one point in Act 2 Scene 2 the title character, echoing Digges' own words, says:

> I could be bounded in a nutshell, and count myself a king of infinite space.

To modern ears that may sound like a glib cliché, but the notion of infinite space was cutting-edge science at the time it was written.

Another person who encountered Digges' ideas was a loud and charismatic Italian named Giordano Bruno, who visited London in the 1580s. While there, he wrote *De l'Infinito Universo e Mondi* ('On the Infinite Universe and Worlds'). As well as espousing Digges' idea of other stars spread throughout the universe, he added another twist – those stars might have planets that are inhabited like our own:

> It is impossible that a rational being, fairly vigilant, can imagine that these innumerable worlds, manifest as like to our own or even more magnificent, should be destitute of similar or even superior inhabitants.

As striking as that thought is to modern readers, it was just one of dozens of radical ideas that Bruno promulgated in his books and public lectures. If there was a common thread to these, it was that Bruno got a kick out of trolling the Catholic Church. Name anything the Church believed, and Bruno was ready to deny it – from the virginity of the Virgin Mary to the divinity of Jesus Christ. It was only a matter of time before the Inquisition got their hands on him, and when they did Bruno's fate was sealed. He was tried for heresy, found guilty and burnt at the stake in February 1600.

There's a danger of misinterpreting the facts here. Giordano Bruno believed in extraterrestrial life, and the Church put him to death in a painful and inhuman way. So does that mean, as some modern writers conclude, that the Church was vehemently opposed to the idea of aliens? Well, no – they were far more concerned by the purely theological heresies that Bruno was spreading. When later, more pious, authors started to talk about extraterrestrials, their existence actually came to be seen as a pro-Christian idea. Why wouldn't an all-powerful God put living beings on all the planets of all the stars in the cosmos?

This picture, which came to be known as the 'plurality of worlds', received a big boost with the publication of a book of that title by the French author Bernard de Fontenelle in 1686. This takes the form of a fictional dialogue in which a philosopher attempts to convince a noblewoman that, among other things, there is life on other planets. He uses a form of reasoning that goes back to the ancient Greek philosopher Aristotle, which argues that things only exist because they were created for a particular purpose. To our eyes, this line of thought looks charmingly naïve:

> You have all the evidence that can be desired: the entire resemblance between the planets and the inhabited Earth; the impossibility of imagining any other use for which they could be created.

Less than a century after Bruno was executed for getting on the wrong side of the church, Fontenelle can be forgiven for taking an unashamedly creationist viewpoint. It was the way most people thought in those days, and it did the trick. *Conversations on the Plurality of Worlds* was an international bestseller, going through more than 100 editions in various languages in its first 100 years. That wasn't the end of its popularity, either. In 1800, a new edition was issued with a preface by the then director of the Paris observatory, who reiterated Fontenelle's argument for a new generation of readers:

> The resemblance between the Earth and the other planets is so striking, that if we allow the Earth to have been formed for habitation, we cannot deny that the planets were made for the same purpose ... Some timid, superstitious writers have reprobated this system, as contrary to religion: they little knew how to promote the glory of their Creator. If the immensity of his works announce his power, can any idea be more calculated than this to exhibit their magnificence and sublimity?

To anyone subscribing to a religious worldview, arguments like this were very persuasive. If the universe is the deliberate creation of a deity – and imagined to be only a few thousand years old – it's reasonable to suppose that similar lifeforms, of a comparable level of intelligence, were placed on every planet. But the 19th century saw the rise of a new, more scientific, viewpoint – and that changed everything. If the universe is unimaginably old, and if events unfold by a complex interplay of chance and natural laws, then all bets are off.

In my first year at university, I was consigned to a grimy Victorian building called Whewell's Court – its only redeeming feature (as far as I was concerned) lying in the fact that it was next door to the biggest bookshop in Cambridge. It does, however, mean that I've heard of the person the building was named after: the obscure 19th-century philosopher William Whewell. In addition to being the person who, in 1834, coined the word 'scientist', Whewell was one of the first people to argue *against* the notion of extraterrestrial life on scientific grounds. Thinking specifically about the other planets of the Solar System, he pointed out that the conditions there would be hostile to any form of life that we're familiar with. No doubt recalling the fate that befell Giordano Bruno, he could see the irony of the situation:

> It will be a curious … event, if it should now be deemed as blameable to doubt the existence of inhabitants of the planets and stars as, three centuries ago, it was held heretical to teach that doctrine.

Whewell's arguments were undoubtedly valid in the case of human-like creatures, which are what the 'God created life everywhere' people were talking about. But much simpler organisms – such as the microscopic bacteria that were only just beginning to be studied in Whewell's time – are a different matter. Another idea that emerged, and briefly flourished, in Victorian times was that simple lifeforms originated in outer space, and arrived on Earth inside meteorites – a theory called 'panspermia'. A quote from the physicist Lord Kelvin, speaking at a meeting in Edinburgh in 1871, summarises the argument succinctly:

> We must regard it as probable in the highest degree that there are countless seed-bearing meteoric stones moving about through space. If at the present instance no life existed upon this Earth, one such stone falling upon it might, by what we blindly call natural causes, lead to its becoming covered with vegetation.

The idea of plant spores, or other fully developed lifeforms, existing inside comets or asteroids remains a possibility, but a very remote one. It attained a certain degree of notoriety in the 1970s, when it was championed by maverick scientists Fred Hoyle and Chandra Wickramasinghe. In their book *Lifecloud* (1978), for example, they argued that 'life arrived eventually on Earth by being showered as already living cells from comet-type bodies'. Although they were part of the scientific establishment – Hoyle at Cambridge and Wickramasinghe at the University of Wales – their views on the topic were far from mainstream, and panspermia remains a fringe theory (to use a polite term).

Having said that, it's now quite obvious – based on direct observation by spacecraft – that asteroids and comets do contain, if not life, then a whole range of complex carbon-based chemicals that might be considered the precursors of life. It's even possible that they carry such materials from one solar system to another.

Interstellar Distances

As far as we know, the universe isn't literally 'infinite', but it's very big. The distances involved are mind-boggling when expressed in kilometres, but there's a more useful unit of distance called a light year. This is the distance that light (or any other form of electromagnetic radiation, from radio waves to gamma rays) travels in a year – about ten10 trillion kilometres.

The nearest star to the Sun, Proxima Centauri, is about 4 light years away. Anyone can picture light whizzing along for 4 years, even if they can't imagine all those trillions of kilometres. There's even a practical consequence, because it means we see the star as it was 4 years ago. Most of the stars easily visible to the naked eye are within a few hundred light years – so the light we see from them set out on its journey after Giordano Bruno went to the stake.

Now let's widen our perspectives. Most discussions of astrobiology talk about life 'in the galaxy', meaning the system of a hundred billion stars of which our Sun is one, and of which the ones we see at night comprise just a tiny fraction. Our Milky Way galaxy is shaped like a disc, roughly 100,000 light years across, with the highest concentration of stars at its centre, about 26,000 light years from us. When astronomers look at the galactic centre through a telescope, they see it as it was around the time humans on Earth had just learned to make pottery.

That's not the end of story, because there are other galaxies – as many of them as there are stars in our own galaxy. But even the closest is over 2 million light years away – meaning we see it as it was before *Homo sapiens* appeared on Earth – and most of them are much further away than that. So there's no hope of any meaningful two-way interaction – which is why astrobiologists tend to focus on our own galaxy.

The Diversity of Life

One of the key realisations that came with the shift from a religious perspective to a scientific one was the idea that organisms take a form that is adapted to their environment, rather than simply being the whim of a creator. The significance of this view for extraterrestrial life was put into words by the French physicist Pierre-Simon Laplace, around the beginning of the 19th century:

> Man, formed for the temperature which he enjoys upon the Earth, could not, according to all appearance, live upon the other planets; but ought there not to be a diversity of organisation suited to the various temperatures of the globes of this universe? If the difference of elements and climates alone causes such variety in the production of the Earth, how infinitely diversified must be the production of the planets and their satellites?

Two hundred years later, pretty much the same sentiment was expressed by Jack Cohen and Ian Stewart in their 2002 book *What Does a Martian Look Like?*:

> Instead of looking for carbon copies of Earth, then, we ought to be theorising about and looking for the different kinds of planets, and other potential habitats for life, that exist out there in the wide universe. 'Exotic' habitats should not be seen as obstacles, but as opportunities; instead of dismissing them with an airy wave of the hand and saying 'obviously life couldn't exist there', we ought to be asking 'what would it have to be like if it did?'

Even here on Earth, there are environments where the familiar forms of life – including ourselves – wouldn't survive for a second. There's the sunless, freezing cold water hundreds of metres below the Antarctic ice, or the super-heated, sulphurous volcanic vents on the deep ocean floor, to name just two. But a close examination of such places almost always reveals colonies of microorganisms – which not only manage to survive there, but positively thrive. Organisms that love such extreme conditions are called, logically enough, extremophiles – and we'll say more about them in Chapter 5.

Despite the enormous diversity of life on Earth, it all belongs to the same evolutionary tree. Species that are close to each other on that tree may look different, because they're suited to dissimilar environments, but they still have underlying features in common. If you say 'animal' to someone, they're most likely to think of a land vertebrate – anything from a frog to an elephant – yet that's really only a small branch of the tree of life (the same one that includes ourselves, for that matter). Despite the wide range of ecological niches, all those animals have the same basic body plan – with four limbs, a central spine, and a head containing a brain and sensory organs. It's highly unlikely that evolution would produce exactly the same body plan on another planet, as Cohen and Stewart point out:

> We would be very surprised indeed to find any aliens with our suite of such immediately understandable terrestrial land vertebrate characters. So surprised, indeed, that we would be sure that they must originally have been terrestrial vertebrates, sharing our ancestry.

That's exactly what UFO believers don't want to hear – because when witnesses describe encounters with aliens, they're always distinctly humanoid in form. A counter-argument, put forward by the more scientifically savvy ufologists, invokes 'convergent evolution'. This is an established principle by which different organisms, adapting to similar niches, may end up independently developing similar forms. Birds and insects, for example – on different branches of the evolutionary tree for more than half a billion years – both use flapping wings to fly. Convergent evolution can lead to striking resemblances, as was the case with an Australian marsupial called the thylacine. Before they died out in the 20th century, thylacines filled a similar niche to wild dogs – and they looked a lot more like dogs than like kangaroos, despite being more closely related to the latter.

It's true that the thylacine was no less dog-like than the aliens reported by UFO witnesses are human-like, but the analogy is a misleading one. Thylacines and dogs have a common ancestor, in the form of the early mammals of the Jurassic era. As ancient as that sounds to us, it's 95 per

**The thylacine was a marsupial closely
related to the kangaroo, yet superficially it
bore a greater resemblance to a dog.**

(Public domain image)

cent of the way from the origin of life on Earth to the present day. The situation would be different if we were talking about a species – the octopus, say – which branched off at an earlier point in the history of evolution. In a marine environment, the octopus fills a somewhat dog-like niche – it's an intelligent predator – so you can imagine its far-future descendants eventually emerging from the ocean to fill a dog-like niche on land. But the land-octopus would still be an eight-limbed invertebrate, and so nowhere near as dog-like as the thylacine.

Another problem with convergent evolution as a justification for humanoid aliens is that all the examples we know of are adapted to life here on Earth. Real aliens would have evolved to fit a completely different planet, differing from ours in surface gravity, atmospheric composition, length of day and a host of other factors. No amount of convergent evolution is going to result in anything that looks remotely like ourselves.

To focus on just one example, consider a planet where the surface gravity is twice that of the Earth. In order to overcome that increased gravitational pull, a creature on that planet has to be stronger, in relation to its mass, than a similar creature on Earth. Paradoxically, that means it has to be smaller. The reason lies in something called the square–cube law – which, I'm afraid, is one of those boring mathematical formulas that give physics a bad name. You may have come across it already, but if not I'll try to explain it as quickly and painlessly as possible.

Imagine you have two solid cubes made of the same material, the first a metre on each side and the second two metres. The volume of the first cube is one cubic metre, and that of the second is eight cubic metres. Because they're made of the same material (and, for the benefit of pedantic physicists, because they're both in the same gravitational field), that means the larger cube weighs eight times as much as the first. But now look at the area of the bottom side of the bigger cube. It's four square metres – only four times larger than the base area of the small cube. So the physical stress* on that side has doubled, because eight times the weight is pushing down through just four times the surface area.

An analogous (but more complicated) logic applies to living creatures as well as cubes. The bottom line is that as their size increases, so do all the physical stresses they

* 'Stress', in a technical sense, is force divided by the surface area it acts on.

have to contend with. That's why ants can get away with flimsily thin legs, and still carry many times their own body weight – but elephants can't. The same reasoning allows us to predict how things would be on a larger planet, with a stronger gravitational pull than that of the Earth. Running through the sums, it turns out that doubling the surface gravity would mean that, for a creature to have the same relative strength and agility as a human, it would have to be half the height and an eighth of the mass. So, as counter-intuitive as it sounds, a bigger, more massive planet is likely to have smaller, less massive inhabitants.

As mentioned in the previous chapter, there are bound to be many planetary systems in the galaxy that are much older than ours, so life may have been evolving there for millions of years longer than on Earth. We can only speculate about what that would mean, but one approach is to consider our own possible futures.

It's a big subject, and one that everyone has opinions on. In very broad terms we can divide the options into three categories. First there's the pessimistic view: that sooner or later the human species will be wiped out by a global catastrophe – either natural or self-inflicted – and it will pretty much be 'back to square one' for life on Earth.

The second option is the optimistic one: the standard 'utopia' scenario, with people living a peaceful, spiritual existence here on Earth, having turned their backs on technology and materialism. As idyllic as this would be for the people involved, it's bad news from an astrobiological point of view, if the aliens took the same route. On a distant planet, they'd be no more detectable to us than microbes.

The third option is to push ahead with technological progress and see where it takes us. This would almost certainly mean the transformation of our species into something else, either via genetic engineering, or by uploading ourselves into machines, or – essentially a combination of the two – by becoming cyborgs. Some people, including Britain's Astronomer Royal, Lord Rees, believe that a 'post-biological' future of this kind is inevitable. One reason is that it's much easier for a suitably designed machine to travel long distances in space than a flesh-and-blood human. To quote Rees on the subject:

> Humans are not the apex of evolution; if we trigger the transition to potentially immortal entities, our role may still be of special cosmic significance … As post-humans journey into space and transcend the limitations of biological entities, our legacy will be their influence in the wider cosmos.

As far as our own future is concerned, it's debatable whether this would be a good thing or a bad thing. But this isn't a book about philosophy, it's about astrobiology. In that context, we'll have to hope that at least some aliens chose to follow this route – because, as we'll see in later chapters, it greatly increases our chances of detecting them.

Life, but Not as We Know It

If the scientific view of aliens makes one thing clear, it's that they won't look anything like us – despite the preponderance of that view in popular culture. This isn't a new observation. As long ago as 1670 Christiaan Huygens, the discoverer of Saturn's moon Titan, wrote:

> 'Tis a very ridiculous opinion, that the common people have got, that 'tis impossible a rational soul should dwell in any other shape than ours. This can proceed from nothing but the weakness, ignorance, and prejudice of men.

Even to this day, the best-known aliens of science fiction – for example the Vulcans, Romulans and Klingons of *Star Trek* – do indeed resemble us to a high degree. And they're not just human-like in outward appearance, but in their behaviour, sensory apparatus, thought processes – even their body language and facial expressions. One reason is obvious: these particular species were all introduced in the 1960s, before there was any such thing as CGI, so they had to be played by human actors in costumes and make-up. But there's another reason, which applies just as much to written works – such as Edgar Rice Burroughs' *A Princess of Mars* and its many sequels and imitators – as it does to on-screen aliens: science fiction isn't science fact. It's a form of entertainment aimed at a human audience, and it has to be appealing and readily understandable to that audience.

But there's another way that sci-fi can maintain a human-interest angle, and that's by depicting the reactions of Earth people when confronted by aliens that are really alien – not just humans with pointy ears or whatever. A few authors even go to the opposite extreme, by depicting lifeforms that are totally different from anything that has ever existed – or could exist – on Earth. A notable example is Fred Hoyle's 1957 novel *The Black Cloud*. At the time he wrote it, Hoyle was working as a university lecturer in astrophysics, and its ostensible subject is a real astrophysical phenomenon: an interstellar cloud of gas and dust. But this one happens to be sentient, and even capable of communicating with people on Earth via radio waves.

The idea isn't totally preposterous. Interstellar clouds are known to contain many complex molecules, and it's just conceivable they could achieve a sufficient degree of organisation for 'thought processes' to occur. It's extremely unlikely, though, because interstellar space is very cold and the gas densities there are very low. On the other hand, anyone looking for 'life not as we know it' could do a lot worse than Hoyle's Black Cloud.

One common – maybe even hackneyed – idea, when sci-fi authors want to talk about 'different biology', is that of silicon-based life. The term is rather confusing these days, because of the prevalence of silicon components in electronic circuits, but that's a different thing altogether. The practical usefulness of silicon stems from its unusual electrical behaviour, while the idea of silicon biology comes from its chemical similarity to carbon. Carbon and silicon are extremely common elements, both here on Earth and in the universe as a whole, but it's carbon that forms the basis for the complex molecules of terrestrial biology. But – so the argument goes – silicon is theoretically capable of forming equally complex molecules, so how about the exotic prospect of silicon-based life?

As it happens, there's plenty of silicon in the Earth's crust – but it shows no interest whatsoever in forming complex molecules. It combines with oxygen to form silica – the primary constituent of most rocks – and seems happy enough to stop there. Out in space, the same is true of all the rocky asteroids we've examined. On the other hand, when carbon is found in asteroids it's often in the form of more complex molecules. So it's difficult to understand why, somewhere else in the galaxy, the situation would be different, and life would emerge from silicon instead of carbon. But that hasn't stopped people from speculating.

The first sci-fi story to feature silicon life was probably Stanley G. Weinbaum's 'A Martian Odyssey' from 1934 ('The beast was made of silica … this thing lived by a different set of chemical reactions; it was silicon life'). But the idea came to the attention of a much wider audience in 1967, with the first-season *Star Trek* episode 'The Devil in the Dark'. That's the one where Dr McCoy, when asked to operate on the rock-like Horta, says 'I'm a doctor, not a bricklayer'.

While we're on the subject of famous *Star Trek* quotations, one that would be bang on topic in the present context is 'It's life, Jim, but not as we know it'. Unfortunately, those words were never actually uttered in the TV series. They originated in the 1987 one-hit wonder 'Star Trekkin'', by British band The Firm. Nevertheless, *Star Trek* did do a lot to popularise the idea of 'life not as we know it'. Another example that cropped up from time to time – perhaps the most far-fetched of all – was the idea of 'pure energy beings'. These made their first appearance in the next episode after 'The Devil in the Dark', called 'Errand of Mercy'. The aliens here, the Organians, were – in the memorable words of Mr Spock – 'as far above ourselves as we are above an amoeba'.

But what does 'pure energy being' mean? In a scientific sense, there's no such thing as 'pure energy', because energy is a property of matter and radiation. It's as meaningless as saying 'pure temperature' or 'pure wavelength'. Presumably what is meant is some form of highly organised data structure – call it 'consciousness' if you're mystically minded, or a complex computer program if you aren't – that can exist without being permanently tied to a chunk of matter such as a human brain. Maybe it's some kind of information-rich modulated signal that can impose itself on whatever pattern of random noise happens to be handy. Okay, I got that idea from a video game,* but that doesn't necessarily mean it's impossible. It's no more likely than Hoyle's sentient cloud of gas, though.

From an astrobiological perspective, perhaps the most interesting aliens in the *Star Trek* franchise are the Borg: part-organic, part-machine entities with a single collective consciousness, capable of rapid propagation through the assimilation of other species. At first sight, that seems as way-out as silicon lifeforms or pure energy beings, and the humanoid appearance of the Borg is completely unrealistic. But the basic vision of the way an advanced spacefaring species might function is worth taking seriously. Particularly notable is their cold, mechanical efficiency, with no human-like emotions – either positive or negative – just an all-out commitment to the survival and growth of the species. That's exactly how it works with lower lifeforms on Earth – and who's to say it isn't the galactic norm?

---

* *1953: KGB Unleashed* by Phantomery Interactive.

But we're straying into philosophical speculation, which we want to avoid in this book. The beauty of the scientific approach is that it doesn't need to get hung up on deep questions in order to make progress. In the case of super-intelligent advanced aliens, for example, it doesn't matter if they're humanoid or octopoid or silicon-based or sentient gas clouds. The way we search for evidence of them isn't going to change. Because we're talking about detecting them at a great distance, not meeting them face to face, it doesn't even matter if they're benevolent or evil or coldly machine-like. The way they generate detectable signals will be the same – and that's what we're going to look at now.

EXTRATERRESTRIAL COMMUNICATION 3

The first person to speculate in a scientific way about the possibility of communicating with extraterrestrials was the great German mathematician Carl Friedrich Gauss. In a letter to the astronomer Heinrich Olbers in 1822, he suggested setting up an array of a hundred huge mirrors to reflect sunlight towards the Moon. This would create a spot of light bright enough to be seen by the Moon's inhabitants, if there were any – who might then respond in a similar way. As Gauss put it, 'this would be a discovery even greater than that of America, if we could get in touch with our neighbours on the Moon.'

By the end of the 19th century, when better observations had shown the surface of the Moon to be totally lifeless, the focus of attention shifted to Mars. In 1892, the French astronomer and science populariser Camille Flammarion proposed that huge geometric shapes, created in the terrestrial landscape, would be visible to Martians – assuming that they existed and possessed telescopes as powerful as our own. At a time of rapid technological progress, this seemed perfectly sensible to Flammarion: 'The idea in itself is not at all absurd, and it is, perhaps, less bold than those of the telephone or the phonograph.'

As Flammarion wrote those words, an Italian inventor named Guglielmo Marconi was putting the finishing touches on a far more effective form of long-distance communication: the radio. Within a few decades, the airwaves would be full of messages flashing around the world – but in the early days, there were hardly any radio transmitters on Earth. In 1899, for example, anyone tuning in to a radio receiver would have heard little more than random noise, emanating from natural sources like lightning and other atmospheric effects. So it was odd when, that very year, the experimenter Nikola Tesla heard what sounded to him like highly structured signals:

> The thought flashed upon my mind that the disturbances I had observed might be due to an intelligent control. Although I could not decipher their meaning, it was impossible for me to think of them as having been entirely accidental. The feeling is constantly growing on me that I had been the first to hear the greeting of one planet to another.

It's not clear exactly what it was that Tesla heard, though it seems likely in hindsight that his mind was simply 'joining the dots' to create a meaningful pattern in what was really just random noise – the auditory analogue of the Martian 'canals' that several people claimed to have seen around the same time. But whether Tesla picked up a signal from Mars or not, the basic idea that alien radio messages might be detectable to us is a good one. So good, in fact, that it's been given a name – and it's one we met already, in Chapter 1: the Search for Extraterrestrial Intelligence, or SETI.

Tuning in to Aliens

In the early 1980s, my first full-time job (if you can call something that was such fun a 'job') involved creating computer simulations of galaxies at the Kapteyn Astronomical Institute at Groningen in the Netherlands. As a theoretician, I was in the minority; most of the staff were observational astronomers analysing data from the nearby Westerbork radio telescope. That was what made the Kapteyn Institute such a magnet for researchers from around the world. Among the many who were there at same time as me was a young American named Seth Shostak. He's going to feature again later in this chapter, because these days he's the Senior Astronomer at the SETI Institute. In those days, however, Seth's work involved studying the structure of galaxies at radio frequencies. And like all the other radio astronomers at Groningen, he was interested in one frequency in particular.

Seen in visible light, galaxies are made up of billions of stars. But there's more to them than that. Another component, which is so pervasive that it's a better tracer of a galaxy's overall structure than the stars, is neutral hydrogen gas. It's far too cold to produce light at visible wavelengths, but it generates a prominent spectral line at a wavelength of 21 cm. 'Spectral lines' are characteristic signatures produced by different types of atom (see box on page 40), and they'll crop up again later in this book in the context of lines in the visible spectrum. But this particular one, the 21 cm neutral hydrogen line, occurs much further down the electromagnetic spectrum, at a radio frequency of 1420 MHz. And that's the magic frequency that Seth Shostak and his colleagues spent all their time looking at.

Spectral Lines

The concept of spectral lines is one of the most important in astronomy. They're the reason astronomers can say, with a high degree of certainty, what the chemical composition of a distant star is. Chemical elements are made of atoms, and one of the fingerprints of a specific type of element is the set of energy levels occupied by the electrons inside its atoms. Sometimes an electron jumps down from a higher energy level to a lower one, emitting the surplus energy as radiation. Alternatively, it can absorb radiation and use the additional energy to jump up to a higher level. Each of these jumps involves a precise 'quantum' of energy, and hence a specific wavelength of radiation.

When you look at a star through an ordinary telescope, you see a combination of all the visible wavelengths it emits. But it's possible to split the wavelengths up using an instrument called a spectrometer – analogous to the way that a prism splits sunlight into a rainbow-like spectrum – to reveal the characteristic lines produced by the various electron jumps that are going on. Downward jumps produce emission lines, which are brighter than the background spectrum, while upward jumps lead to absorption lines, which are darker. By measuring the exact wavelength of a line, astronomers can work out which particular chemical element produced it.

Many of the most important spectral lines occur in and around the visible part of the spectrum, but some very low-energy jumps result in the longer wavelengths associated with radio waves. That's where the 21 cm neutral hydrogen line fits in.

The 21 cm line was originally recognised as a window onto galactic structure in 1951, by the Dutch astronomer Jan Oort. He's best known today for the 'Oort Cloud' – his notion of a huge repository of comets in the outer Solar System – but in the 1950s he was one of the great pioneers of radio astronomy. He used 21 cm observations to show that our galaxy, like many external ones, has a spiral structure – something that's impossible to discern at visible wavelengths.

You may think we've strayed a long way from the subject of extraterrestrial intelligence – but now we'll come back to it with a bang. In 1959, the journal *Nature* printed a paper by physicists Giuseppe Cocconi and Philip Morrison entitled 'Searching for Interstellar Communications'. They made the point that, if aliens wanted to set up an interstellar radio beacon to attract the attention of other civilisations, the best frequency to use would be 1420 MHz. We've already seen that Earth-bound radio astronomers spend a lot of time making neutral hydrogen observations at that frequency, and the same is almost certainly true of radio astronomers throughout the universe.

I may have laboured this point a little, but that's because I'm irritated by the way other writers gloss over it – leaving the reader with the impression there's something arbitrary or even mystical about the way SETI scientists focus on that 21 cm, 1420 MHz window. One book, for example, simply states that 21 and 1420 'were guessed to be numbers with universal cosmic significance'. Wrong! The logic is much cleverer and stronger than that. To do their day job properly, radio astronomers need to spend a lot of time looking at the 21 cm hydrogen line – not just here and there, but creating whole maps of the sky at that wavelength. Any aliens who know about radio waves, and are interested in the universe beyond their own planet (and if you think about it, both those things are prerequisites for any civilisation that's going to send out a deliberate interstellar radio message) will be equally aware of the importance of the 21 cm wavelength. If they want to maximise the chances of their beacon being seen, that's the place to put it.

The first person to put SETI into practice was Frank Drake at the Green Bank radio observatory in West Virginia. At a hundred metres in diameter, Green Bank is the largest steerable radio dish in the world – a third as big again as Jodrell Bank's iconic Lovell telescope.

In 1960, Drake pointed that giant dish at two nearby Sun-like stars – Tau Ceti and Epsilon Eridani – and listened on that critical frequency of 1420 MHz. He didn't hear anything from either of those stars, but at one point while he was slewing the telescope from one direction to another he did pick up a coded signal coming from the sky. The source was travelling so rapidly, and at such enormous altitude, that Drake realised it was either an extraterrestrial UFO or a high-tech aircraft with capabilities beyond anything in the public domain. It turned out to be the latter. A few weeks later, CIA pilot Gary Powers was shot down over the Soviet Union in his super-secret U-2 spyplane – and Drake belatedly put two and two together.

As it happens, Frank Drake's most important contribution to SETI wasn't practical but theoretical – in the form of the Drake equation, which he formulated in 1961:

$$N = R^* \times f_p \times n_e \times f_l \times f_i \times f_c \times L$$

This is an impressive string of mathematical symbols, the meaning of which will become clear in due course, and it's often emblazoned on mugs and T-shirts by enthusiasts who vaguely imagine that it somehow 'proves aliens exist'. At the other end of the spectrum, people who take the time to think through what each of the symbols means often come to the conclusion that it's meaningless nonsense. But both views miss the point of what Drake was trying to achieve with his 'equation'.

Part of the problem is that it's not really an equation. An equation is an exact mathematical relationship, such as Newton's second law of motion, which says that force equals mass times acceleration. If you plug in numerical values for the mass and acceleration of an object, what comes out is the precise numerical value of the force acting on it. Drake's equation is something else entirely – a simple intuitive formula designed to provide a rough numerical estimate of an unknown quantity. That's the sort of thing scientists and engineers use all the time, though they don't generally dignify it with the name 'equation'. It's simply a sanity check to work out if there's any merit at all in a particular assertion.

Another area of confusion concerns the specific assertion the Drake equation is designed to test. It's not the big philosophical one, 'the galaxy is full of intelligent alien civilisations', but a much more practical one: 'it's worth investing radio telescope observing time in listening for messages from aliens'. The answer doesn't even have to be 'yes' to make the effort of listening worthwhile – just so long as it isn't an emphatic 'no'.

To see how the Drake equation works, let's apply the same line of reasoning to a simpler question. How many professional astronomers are there in the United States? Our starting point is a statistic that's easy to look up: the average number of astronomy PhDs awarded by US universities in a year, which turns out to be something like 150. Not all those PhD recipients will go on to become professional astronomers, but let's make the reasonable-sounding assumption that half of them do. An average professional career lasts around 40 years, so multiplying those three numbers together we get $150 \times 0.5 \times 40 = 3,000$.

That's our estimate of the number of professional astronomers in the United States. It's unlikely to be the correct answer, but it's going to be in the right ballpark. We can be pretty sure the answer isn't just 'a dozen or so', and it isn't 'millions of them'.* The aim was to get a feel for the answer – and it's exactly the same with the Drake equation.

---

* According to Wikipedia the American Astronomical Society has approximately 7,000 members – but not all of them will be professional astronomers.

Of course, Drake's aim was a much more challenging one: to estimate the number of alien civilisations that might be sending out radio signals. The basic logic is exactly the same, though. You start with the number of new stars formed in the Milky Way galaxy each year, multiply by the fraction of those stars likely to have radio-transmitting civilisations in their vicinity (actually Drake breaks this down into a long string of intermediate factors, as shown in the table below), and finally multiply by the number of years such a civilisation remains interested in radio transmission.

Frank Drake's estimate of the number of radio-transmitting civilisations

Factor	Drake's guess at lower bound	Drake's guess at upper bound
New stars formed each year in the galaxy (R*)	1	1
Fraction of stars with planetary systems (f_p)	0.2	0.5
Number of habitable planets in each system (n_e)	1	5
Fraction of habitable planets with life (f_l)	1	1
Chance that life develops intelligent civilisation (f_i)	1	1
Fraction of civilisations transmitting radio (f_c)	0.1	0.2
Number of years such a civilisation lasts (L)	1,000	100,000,000
Multiplied together	20	50,000,000

The first factor in the Drake equation, the number of new stars formed in the galaxy per year, is the only one that's reasonably well established. It's actually slightly larger than Drake's figure of 1, but that's still a good round-number estimate. The other factors are subject to debate, and the current consensus would put some of them slightly higher, and several of them a lot lower, than Drake's guesses. The final number – the length of time that a radio-transmitting civilisation persists – is particularly controversial. This isn't simply the expected lifetime of an intelligent species – in which case Drake's upper estimate of a hundred million years might be just about credible – but the length of time the species perseveres with its project of blasting out interstellar radio signals. Viewed in that way, even Drake's lower limit of a thousand years looks distinctly optimistic.

The fact that most of the factors in the Drake equation are subject to uncertainty is the reason its critics hate it so much. But they're missing the point. The purpose of the exercise wasn't to pin down a precise number, but to show that SETI isn't a complete waste of time. The really disastrous result would be if there was no uncertainty at all – and the final answer came out as a minuscule number, much less than one. That would mean that SETI had no hope of success at all.

As it is, Drake's suggested 'lower estimate' of 20 can probably be taken as a sensible 'upper estimate'. It's still high enough to make SETI worth the effort – especially in light of the huge payoff if it turns out to be successful. That's a conclusion many people have come to over the years, anyway – to the extent that radio-based SETI has become one of the central pillars of observational astrobiology.

60 Years of SETI

In the six decades since Frank Drake pointed the Green Bank telescope at Tau Ceti and Epsilon Eridani, SETI has only once turned up something that looked like it might be an intelligent signal from an extraterrestrial source. It was picked up in August 1977 by a different telescope – called Big Ear, at Ohio State University – using the same neutral hydrogen frequency of 1420 MHz. The signal took the form of a sharp spike, rising way above the background noise as the telescope scanned across the sky, and was exactly the sort of thing the researchers were looking for. When one of them, Jerry Ehman, spotted it on the printout, he wrote 'Wow!' next to it – and the name stuck.

The picture below shows a graphical representation of the Wow signal, with time along the horizontal axis and intensity on the vertical axis. Back in the days when

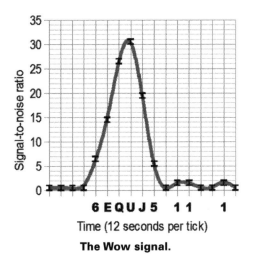

The Wow signal.

(Wikimedia Commons user Cmglee, CC-BY-SA-3.0)

computer results were printed out on paper, that would have been a very wasteful way to output a large amount of data. Instead, the Big Ear telescope simply churned out a long alphanumeric sequence – one character per time sample – with each character measuring the ratio of the signal strength to the background noise level. From 0 to 9 the digits meant exactly what they said, after which the letters A to Z represented numbers from 10 up to 35.

In this system, the Wow signal came out as '6EQUJ5' – representing a sharp rise up to a peak at $U = 30$, and then a sharp drop back to the background level. That's unfortunate in a way, because it makes the signal look, at first glance, like a coded message (and a search through the dodgier internet forums will turn up plenty of people claiming to have decoded it). But it's not really a message at all, just a momentary blip – and one that was never seen again.

No one knows what produced the Wow signal. It might have been aliens, or it might have a completely different explanation. One suggestion, made by astronomers Antonio Paris and Evan Davies in 2016, is that Big Ear detected exactly what astronomers expect to see at a frequency of 1420 MHz: neutral hydrogen gas. Why would such a strong, highly localised, source of it be at a particular spot in the sky one day and then never seen again? Because, Paris and Davies say, the signal came from a comet.

This makes sense up to a point, because there happened to be a couple of comets in the sky at that time. But they weren't in the direction that Big Ear was looking. On top of that, Seth Shostak – who, as I've already mentioned, was a neutral hydrogen expert long before he was a SETI expert

– has pointed out that comets don't produce strong hydrogen signals. So – fortunately for those of us hoping the Wow signal really was aliens – there still isn't a convincing natural explanation for it.

From what I said earlier, about radio astronomers spending huge amounts of time looking at the 1420 MHz frequency because they're mapping the distribution of neutral hydrogen, you may wonder why people bother to do dedicated SETI searches at all. Surely if the aliens have created a beacon at that frequency, it would show up in one of those maps? Unfortunately it isn't that simple. Artificial signals are likely to be characterised by a rapid time variation – analogous to a series of clicks or beeps, if you think in audio terms, rather than a long steady tone. But radio astronomers aren't normally looking for rapidly varying signals, so they take the equivalent of a long-exposure photograph – a very long one – to get the best 'picture' possible. That washes out the temporal structure that would be the hallmark of an artificial signal.

One radio astronomer who did make a specific attempt to listen for time-varying signals, back in 1967, was a PhD student at the University of Cambridge named Jocelyn Bell Burnell. She wasn't looking for aliens, but for an effect called interplanetary scintillation, which occurs when radio waves from a distant astronomical object pass through ionised gas in the Solar System. That may not sound very exciting, but it's useful to astronomers because it tells them something about the size of the distant object from which the radio waves are coming. Anyway, that was Jocelyn's PhD project, and so (as anyone who's ever done a student project will understand) she was pretty single-minded about the subject.

The problem was, when she looked in one particular direction at one particular frequency – 81.5 MHz, much lower than the neutral hydrogen window – she found an annoyingly repetitive signal that clearly wasn't interplanetary scintillation. It just beeped away (to use the audio analogue again) roughly once every 1.337 seconds. No natural process was known that could produce a signal like that, and it looked for all the world like an interstellar beacon set up by some distant extraterrestrial civilisation. Jocelyn was far from thrilled by the thought, as she explained a few years later:

> Here was I trying to get a PhD out of a new technique, and some silly lot of little green men had to choose my aerial and my frequency to communicate with us.

That phrase 'little green men' got so tied up with the mysterious source that, for a time, it was actually named 'LGM-1' – although to be honest no one took the alien explanation very seriously. Eventually Jocelyn and her supervisor, Tony Hewish, worked out what was really going on. The radio pulses were coming from a very small, extremely rapidly rotating star. Now known as pulsars, there are over a thousand such stars in the galaxy – but they're all perfectly natural phenomena, and nothing to do with aliens.

Despite the near misses of the Wow signal and LGM-1, astronomers still haven't found definitive evidence for an alien radio beacon. That's disappointing, but perhaps not all that surprising. Why should we expect aliens to go to the trouble of setting up a beacon – which would need a large amount of power over a long period of time – in the first place?

If you go back to the Drake equation, there's an ambiguity over whether it's talking about a deliberate beacon aimed by aliens at other civilisations – in which case the logic leading to the 1420 MHz frequency makes sense – or simply 'domestic' radio transmissions the aliens use for their own purposes. The latter are far more likely to occur, but – because it's a matter of picking up unintentional leakage – harder to detect at a given distance. Added to that, we have no idea what frequencies the aliens are going to use – except that it won't be 1420 MHz. They're going to keep that one as quiet as possible for the benefit of their radio astronomers, just as we do here on Earth.

SETI researchers do look for 'leakage' signals over a wide range of frequencies, but the result is an enormous quantity of data that requires a lot of computer time to sift through. The solution they came up with was SETI@home, originally set up by the University of California in 1999. Computer users around the world volunteer to allow their idle processor time to be used to scour radio telescope data looking for meaningful signals. By the end of its first decade, SETI@home had become the biggest distributed computing project in history, with over 6 million users around the globe. It's also the most successful in terms of the quantity of data analysed – but not, unfortunately, in terms of finding evidence of extraterrestrial intelligence.

That's disappointing – but we have to remember the sheer scale of our galaxy, with something like a hundred billion stars in total. Even if a hundred, or a thousand, of those have civilisations actively transmitting radio signals, it's still like looking for a needle in a haystack. On top of that, signals that have travelled hundreds or thousands of light years are going to be extremely weak by the time they reach us. And if they're leakage from transmissions that were never meant to travel interstellar distances, they won't have been particularly strong to start with.

Fortunately, our own technology is improving all the time, and future radio telescopes will be far more sensitive than they were when Frank Drake first embarked on SETI. The proposed 'Square Kilometre Array' – literally a square kilometre's worth of receiving dishes to be constructed in South Africa and Australia in the 2020s – should be able to detect the equivalent of a terrestrial mobile phone system anywhere within 50 light years of Earth.

Another problem SETI has had in the past is that it has been an assortment of small-scale independent initiatives, rather than a single, highly coordinated one. As authoritative as it sounds, the SETI Institute – where Seth Shostak works – is a small, privately run organisation set up in 1984 under the scientific leadership of Jill Tarter. Its funding comes from various private individuals, with nothing at all from the US government. At the start of 2019, in an attempt to bring all the disparate research strands together, Jill Tarter launched a new web-based tool called Technosearch – essentially a huge database of every single SETI search from 1960 up to the present.

In another promising development, the Russian billionaire Yuri Milner is pouring $100 million into the most ambitious SETI programme to date, called Breakthrough Listen. That money will buy observing time, amounting to thousands of hours a year, on two radio telescopes – the giant Green Bank dish in the USA and the Parkes telescope in Australia

– as well as using optical observatories to search for coded laser transmissions. By the time it comes to an end, in 2025, Breakthrough Listen will have enormously increased the amount of data available to SETI researchers – and for that matter, to members of the public using the SETI@home software.

Meaningful Messages

If and when a SETI message is detected, that will just be the beginning. The next problem is to work out what it says. Assuming it's a deliberate attempt by aliens to communicate with another civilisation, how would they go about making themselves understood? That's a question SETI experts have grappled with over the years, and even tested out a few ideas in messages they've transmitted – rather than received – through radio telescopes.

The most famous of these was the Arecibo message, blasted out through the 300-metre dish antenna at Arecibo observatory in Puerto Rico. That's three times the size of Green Bank – but rather than being a steerable dish, it's built into a natural depression in the landscape. That means it's constrained to look – or transmit – at a narrow cone of angles around whatever point in the sky happens to be overhead at the time.

The Arecibo message was transmitted on 16 November 1974, as part of a ceremony to mark the reopening of the observatory after an upgrade. Like other similar gestures before and since, it wasn't a serious attempt to communicate with aliens – the transmission was much too brief for that.

It wasn't aimed at a nearby star system, either. At the time of the ceremony, the telescope happened to be pointing at a star cluster in the constellation Hercules, an unpromisingly distant 24,000 light years away. But the Arecibo message was really just an exploratory exercise in constructing a message that might be understandable to aliens.

As seemingly impossible as that task is, a few assumptions can be made. If the aliens are using technology similar to our own to receive the message (rather than, say, having a natural ability to perceive radio signals), that implies they have a similar understanding of science and mathematics – without which they couldn't design and build the necessary technology. Based on this reasoning, the Arecibo message was all about science and mathematics. Encoded in the form of binary arithmetic – the closest thing to a 'natural' system of mathematics that we know – it contained brief descriptions of DNA, human beings, the Solar System and the Arecibo telescope itself.

If you've got time to waste, you can look up the Arecibo message on Google and see what you make of it. Personally I find it almost completely meaningless, even after I've been told what it's supposed to represent. I doubt the aliens would have a clue what it meant either.

But does that really matter, if the purpose of the message is to tell the aliens we're here? That simply requires a message that looks like a message – rather than the result of a random natural process – regardless of whether the recipients can decipher it or not. As it happens, there's a clever way to tell if a message is written in a properly formed language, without needing to know what it actually says.

Called Zipf's law, it's based on the frequency of occurrence of the various symbols in a message. All human languages, for example, obey Zipf's law, as does the 'speech' of bottlenose dolphins and humpback whales. But the empty chattering of squirrel monkeys doesn't obey it – and neither, amusingly enough, does the Arecibo message.

Never mind – if nothing else, the Arecibo message was a useful exercise in science outreach, bringing the subject of SETI to the attention of a wider audience. The same is true of another 'message to extraterrestrials' from the same decade: the gold-plated plaques carried by NASA's Pioneer 10 and 11 spacecraft, which are now heading out of the Solar System. They're not doing so very quickly, though, by interstellar standards. It will be 4 million years before one of them passes anywhere near another star system – in the constellation Aquila – and even then its chances of being intercepted by aliens are vanishingly small.

One of the people involved in designing the plaques was the well-known astronomer Carl Sagan. Soon after the two spacecraft were launched, he wrote about the subject in his book *The Cosmic Connection*.* Paradoxically, one chapter is called 'A Message to Earth'. That looks wrong, but it's what he meant – as he emphasises at the end of the chapter:

> The greatest significance of the Pioneer plaque is not as a message to out there; it is a message to back here.

---

* Having dug my copy out, I see there's a sticker in the front: 'North Bromsgrove High School Prize – awarded to Andrew May for Fifth Form Physics, 1974'. So you can see how long I've been interested in this subject!

The picture below shows a sketch of the plaque. The left-hand side – the sciencey bit – provides a symbolic depiction of the Earth's location in the Solar System (the diagram at the bottom roughly translates as 'third rock from the Sun') and the Solar System's location in the galaxy (the lines indicate the directions and distances to some nearby pulsars, with the frequency of the pulses shown in binary notation). All this might be of interest to aliens, but it's also a sharp reminder to the plaque's human audience of their insignificant place in the cosmos.

On the right-hand side – and unlikely to be of interest to anyone in the universe except ourselves – are line drawings of male and female human figures. Quite apart from the 'four-limbed vertebrate' shape that is probably unique to

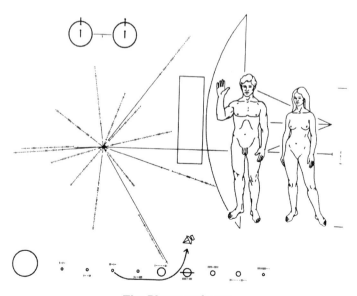

The Pioneer plaque.
(NASA image)

Earth, the very form of the drawings may, as Sagan points out, be indecipherable to extraterrestrials. Even if they have optical sense-organs capable of seeing the image, they may have a completely different way of depicting three-dimensional objects on a two-dimensional surface.

So we've established that the Pioneer plaque was basically an exercise in public relations. Even in that context, however, it was far from being an unqualified success. Sagan received letters complaining about NASA 'sending those dirty pictures of naked people out into space'. Others were worried that the female looks more passive – and by implication less important – than the male. It's also true that, largely on the basis of their hairstyles, both figures come across as ethnically Caucasian. But that wasn't the intention, according to Sagan. The original sketches, by his wife, showed the man with an afro haircut and the woman with dark hair – but those nuances got lost when the final engraving was made.

The unfortunate result is that, as the world has moved on and issues of diversity have become increasingly important, the Pioneer plaque hasn't aged well. That's a very human perspective, of course; any aliens intercepting the Pioneer spacecraft are unlikely to be offended by what Dr Jill Stuart – an expert in space policy at the London School of Economics – described in 2015 as 'Western-dominated material'. On the other hand, those same aliens might at a pinch manage to decode the map showing Earth's location in the galaxy. And that might or might not be a good thing.

Back in 1983, during my time there, the Kapteyn Institute hosted a conference on galactic astronomy – and one of the after-dinner speakers was Seth Shostak. The subject he chose to talk about was 'Life in the Galaxy' – treated, as befits the occasion, in a generally light-hearted way.* Even so it was a pretty good overview of the subject, covering (I'm looking at the conference proceedings here, not quoting from memory) the Fermi paradox, the Drake equation, biosignature gases in planetary atmospheres and even the possibility of post-biological evolution. Seth also discussed the Pioneer plaque:

> Even assuming the message is retrieved and correctly decoded … we might have the wretched luck to make our presence known to an aggressive culture which will destroy civilisation to obtain our chlorophyll or other commodity. In this view Carl Sagan may be destined for eternal fame as the man responsible for the obliteration of Earth. But a hundred thousand years is a long time, and Carl isn't worried.

The same argument could be – and has been – made against sending out Arecibo-style messages. Because these travel at the speed of light, they could potentially be intercepted in a much shorter timeframe. The resulting 'danger' was brought to public awareness by no less a personage than the physicist Stephen Hawking, who wrote in 2010 that:

> If aliens visit us, the outcome would be much as when Columbus landed in America, which didn't turn out well for the Native Americans.

---

* When I mentioned the talk to Seth a few years ago, he told me he's always considered it to have been the first public presentation he gave on the subject of SETI. So I was present at a historic event, even though I didn't realise it at the time.

As striking as that analogy is, it's not a good one. Columbus didn't travel to the New World because its occupants had advertised their existence; he was searching for natural resources. That's one possible reason why aliens might come to Earth, too – although they'd probably find it more cost-effective to pillage other parts of the Solar System instead. Rare elements would be far easier to extract from small asteroids than from the Earth, while water – if that's what they're after – is far more plentiful, in the form of ice, in the outer Solar System than it is on our own planet. In any case, whether it's a likely state of affairs or not, resource-pillaging aliens – just like Columbus – are going to come whether or not we signal our presence with radio messages.

One person who has spoken out against Hawking's 'keep a low profile' view – not once but on several occasions – is Seth Shostak, in his current role as one of the principal spokespeople of the SETI Institute. His argument is that there's no point trying to hide our existence from aliens, because if they're advanced enough to travel between the stars they'll have the technology to detect our presence anyway. Here's what he said to the *New York Times* in 2015:

> The inability to gauge this peril prompts some critics to argue that ... we should simply forbid powerful transmissions to the skies. Indeed, a small consortium of academics in California has drafted a petition urging this. It's a wary approach. It's also poor insurance. Any extraterrestrials with technology advanced enough to threaten us will surely have antennas larger than our own, instruments

that can pick up the television and radio signals broadcast willy-nilly since World War Two. We are already shouting into the jungle, albeit with less volume than a deliberate signal. But the dangerous creatures may have good hearing.

The plain fact is that without knowing the aliens' motives – or their worldview or concept of morality – there's no way of knowing if attracting their attention is a good thing or a bad thing. Do they want to help us develop and mature, like the Vulcans in *Star Trek*, or are they going to assimilate us into a soulless empire, like the Borg? Either way, the question is academic, because actual face-to-face contact with intelligent aliens is extremely improbable any time soon. It's far more likely that we'd receive a coded message via radio or laser transmission – and there's no danger in that, is there?

Actually there might be. As mentioned in the previous chapter, the astrophysicist Fred Hoyle occasionally dabbled in science fiction – his 1957 novel *The Black Cloud* has already been mentioned. Four years later he collaborated with professional author John Elliot on *A for Andromeda*, in the form of both a novel and a TV series. This is an edge-of-the-seat sci-fi thriller, so much better written than *The Black Cloud* that I suspect it's virtually all Elliot's work, with Hoyle merely supplying the basic idea. It is, however, a brilliantly clever idea.

The protagonists pick up a radio signal from space which, when it's decoded by a computer, turns out to be instructions for building another, much more powerful computer. This in turn shows the scientists how to create a new form of life – and before they know it, the alien takeover has begun. Almost exactly the same premise, updated for a DNA-savvy audience, can be found in the 1995 Hollywood movie *Species*.

Of all the alien threats in science fiction, this may be the most credible. I don't mean the fact that, in both *A for Andromeda* and *Species*, the alien invader takes human form – which is decidedly improbable – but simply the idea that a message could take the form of a malicious computer program. When Hoyle and Elliot wrote their novel in 1961, the idea of using a computer to decode the alien message was a novelty. The standard, in those days, would be to record the signal on analogue tape and then have it analysed by human linguists and cryptographers. But that's all changed now, and the message would be recorded in digital form before human scientists were even aware it had been received. The idea of using anything but a computer to do the hard work of analysing and interpreting it is all but unthinkable.

It's a sad fact of modern life that 'malicious software' is all too common. So why shouldn't aliens have the same idea? Not all of them, of course, but – as in the world of computer programmers – just a small but dangerous minority? Here's what astronomers Michael Hippke and John Learned said on the subject in a 2018 paper:

> While it has been argued that sustainable extraterrestrial intelligence (ETI) is unlikely to be harmful, we cannot exclude this possibility. After all, it is cheaper for ETI to send a malicious message to eradicate humans compared to sending battleships. If ETI exist, there will be a plurality of good and bad civilisations. Perhaps there are few bad

ETI, but we cannot know for sure the intentions of the senders of a message.

One possibility suggested by Hippke and Learned, if the senders of the message have intercepted our own domestic communications and know something about our worries and insecurities, is that they might try to sow fear with spurious threats like 'we will make your sun go supernova'. Or they could troll us with deliberately divisive statements, such as claiming that one of our religions is true and all the others are false. Messages like that could cause almost as much trouble as an invasion fleet.

Another possible threat discussed by Hippke and Learned – and closer to the *A for Andromeda* scenario – is a long coded message prefixed with a statement along the following lines:

> We are friends. The galactic library is attached. It is in the form of an AI which quickly learns your language and will answer your questions. You may execute the code following these instructions …

If you saw that in an unsolicited email, you'd delete the message without opening the attachment. Sadly, that's what Hippke and Learned conclude would be the wise thing to do in the case of an alien message too. All in all, it might be safer if we could find a way to observe the aliens without actually communicating with them – and that's what the next chapter is about.

INTERSTELLAR ENGINEERING

4

Even if there are thousands of intelligent civilisations in the galaxy, only a small minority of them – if any – are likely to be blasting out radio messages we could detect. If the messages are aimed at us, or other civilisations like us, the senders would have to share our interest in interstellar communication – which may not be the case. On the other hand, if we're just talking about accidental leakage from domestic communications, it implies the aliens are going about it in an energy-inefficient way – which seems unlikely for an advanced civilisation.

It's far more likely that any aliens that might exist are simply minding their own business. Without knowing the exact nature of that business, there's still a chance that it would produce technosignatures – tell-tale signs that the aliens are there – that we could detect.

If they're at the same technological level as us, for example, we can start by imagining what unusual features about our own planet might reveal our presence to a distant observer. One possibility is nightglow. Seen from space, the side of the Earth that should be pitch black – the side facing away from the Sun – is artificially illuminated by millions of tiny pinpoints of light, coming from cities and other products of human civilisation. The situation on any other planet with a similar civilisation will be the same, resulting in a characteristic signature in terms of the light emitted. By astronomical standards the light will be very faint, and our current telescopes aren't powerful enough to spot it at a distance measured in light years – but the next generation of instruments might be.

Another tell-tale sign of our presence is something we can be less proud of, and that's industrial pollution in the atmosphere. When we go on to talk about biosignatures – the microbial counterpart of technosignatures – in Chapter 6, we'll see how it's possible to detect chemicals in the atmospheres of exoplanets – planets orbiting other stars. Here on Earth, those biosignatures have been present for billions of years – as long as life has existed. But there are other, much more recent, chemicals in the atmosphere produced by industrial processes. Ironically, the most useful indicator in this category is a particularly harmful group of greenhouse gases called chlorofluorocarbons (CFCs). These are highly unlikely to be produced by any natural process, and should be readily detectable by the James Webb Space Telescope due to be launched in 2021.

As far as generating technosignatures is concerned, our own civilisation is in its infancy – we've been doing it for a few hundred years at most. But cosmic time is measured in billions of years. The chances are that some extraterrestrial civilisations will be, if not billions, then certainly millions of years ahead of us. And that raises some interesting possibilities.

Thinking Big

Imagine a technological civilisation that's millions of years old. It might, as sci-fi authors usually assume, venture out into interstellar space. But what if it stayed at home? One thing that seems inevitable, based on examples here on Earth, is that any successful population will grow and grow. So where will they go when they've filled up their home planet? The most comfortable place will be at the same distance from their sun as that planet. So we can imagine them moving 'building material' from elsewhere in their planetary system to construct a whole swarm of habitats orbiting the central star at exactly that ideal distance.

The end result will be a Dyson sphere – so called after the physicist Freeman Dyson, who proposed the idea in 1960. He was thinking in terms of the population growth problem, but there are other reasons why an advanced civilisation might want to create such a megastructure. It would trap a huge amount of solar energy – far more than normally falls on the surface of a planet – and energy is an essential enabler for a whole range of possible technologies. At the introverted end, maybe the aliens have chosen to live in a vast Matrix-style virtual reality universe of their own creation. At the opposite extreme, maybe the Dyson sphere provides the energy to operate a wormhole to another part of the space–time continuum.

That type of speculation is pure fantasy, of course, and it's not what Dyson's original paper was about. He wasn't really focusing on why the aliens might construct such a sphere, or what could be done with it. He was interested in a much more practical question. If Dyson spheres exist out there in the galaxy, could we detect them?

It turns out that a Dyson sphere would indeed produce a distinctive signature we could detect. The title of that original paper, published in the journal *Science* in June 1960, was 'Search for Artificial Stellar Sources of Infrared Radiation'. The idea is that a huge swarm of artificial objects around a star would absorb a large proportion of the light we

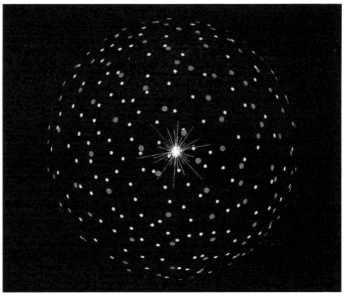

A Dyson sphere: a huge swarm of artificial objects in orbit around a central star.

(Wikimedia Commons user Rfc, CC-BY-SA-3.0)

would normally expect to see from it. Here on Earth, most of the light that falls on the planet's surface is eventually re-radiated into space in the form of heat – or more technically, as infrared waves. It would be exactly the same, albeit on a much vaster scale, with a Dyson sphere. So all we have to do is look for stars emitting an anomalously high proportion of energy in the infrared.

Since Dyson's original suggestion, several observational searches have been made for Dyson spheres. The most comprehensive of these, by Richard Carrigan of Fermilab* near Chicago, used data from IRAS (Infrared Astronomical Satellite), which observed a quarter of a million sources in a survey covering 96 per cent of the sky. Unfortunately, the results were inconclusive, as Carrigan explained:

> This search has shown that at best there are only a few quasi-plausible Dyson sphere signatures ... This limit includes both pure and partial Dyson spheres. With several possible exceptions all the good sources identified in this search have some more conventional explanation other than as a Dyson sphere candidate. In spite of the fact that there are many mimics such as stars in a late dusty phase of their evolution, interesting Dyson sphere candidates are quite rare.

The infrared technosignature predicted by Dyson is implicitly assumed to be constant over time: whenever you look at such a star, you will always see the same reduced visual brightness

---

* Named for Fermi's work on nuclear physics, not the Fermi Paradox.

and increased infrared emission. That would be the case with a classic Dyson sphere, made up of a dense swarm of small objects orbiting the central star. But what about an asymmetric megastructure, which sometimes obscures our line of sight to the star and sometimes doesn't? Such a set-up would still produce a distinctive technosignature – but in this case it will be a time-varying one.

This idea was put forward by the French astronomer Luc Arnold in 2005, in a paper entitled 'Transit Light-Curve Signatures of Artificial Objects'. Later, in Chapter 6, we'll see how the Kepler space telescope was designed to detect exoplanets as they pass in front of their host star. Such a passage is called a transit, and its effect is seen in the star's 'light curve' – a very slight dimming when the planet is in front of it. Arnold's idea was that an artificial megastructure would produce a different, but equally distinctive, light curve signature of its own.

Kepler was launched in 2009, four years after Arnold's paper. The spacecraft captured data from hundreds of thousands of stars, and their light curves were scoured for signs of planetary transits both by professional astronomers and by the citizen science project Planet Hunters. In a coup for the latter, they discovered a peculiar light curve in 2015 that looked nothing like a planet, and very like what Luc Arnold had predicted for an artificial megastructure.

It was associated with a star labelled KIC 8462852 in the Kepler catalogue. Fortunately this was soon given the friendlier name of Tabby's Star, after Tabetha Boyajian, the head of the Planet Hunters initiative, and lead author on the paper announcing the discovery. When the news hit the mass media, it acquired yet a third name: 'the alien mega-structure star' – even though most journalists admitted in the small print that the observed light curve had other possible explanations besides a swarm of artificial objects.

Amusingly, that name persisted even after one of those alternative explanations emerged as the odds-on favourite. Take this January 2018 headline from the *National Geographic*, for example: 'Mystery of Alien Megastructure Star Has Been Cracked'. Reading further, it's clear that it almost certainly isn't a megastructure at all, but a cloud of dust. Recounting further work by Boyajian and her team, the article quotes her as follows:

> If a solid, opaque object like a megastructure was passing in front of the star, it would block out light equally at all colours. This is contrary to what we observe.

Despite this, it may be a long time before the media learn to refer to Tabby's Star as anything but 'the alien megastructure star'. Never mind; the idea hasn't been completely debunked. Although it looks unlikely in this particular case, Arnold's basic argument remains as valid as ever – and light-curve dimming is still a promising way to search for megastructure candidates.

There might be other ways in which an advanced civilisation could produce detectable technosignatures. The one we've looked at so far, megastructures, makes sense if the aliens choose to remain in the vicinity of their home planet. But what if, like the aliens of sci-fi, they decide to venture out into the galaxy? The subject of interstellar spaceships has an uphill battle to be taken seriously, thanks to the lack of credibility of most UFO sightings. But instead of looking at odd things in the sky and saying 'that must be an alien spaceship', let's approach the problem more scientifically. If interstellar space vehicles exist, could we hope to detect them? It turns out the answer is yes – possibly from many light years away.

The key lies in that fact that it requires a tremendous amount of energy to accelerate even a small spacecraft to close to the speed of light – and that sort of speed will be essential if interstellar travel is to be practical. Yet no matter how well designed the propulsion system is, it's unlikely to be 100 per cent efficient. That means a fraction of the energy will go to waste – and might potentially be detectable.

This possibility was discussed by Robert Zubrin at a conference in 1994, in a paper entitled 'Detection of Extraterrestrial Civilisations via the Spectral Signature of Advanced Interstellar Spacecraft'. Without worrying about the engineering details, he considered a range of basic physical principles that might underlie a hypothetical 'starship drive' technology. His conclusion was that each of these would generate a characteristic form of radiation – ranging from low-frequency radio waves all the way up to gamma rays, depending on the physics involved – which might be detectable over interstellar distances using standard astronomical instruments.

The problem is that similar types of radiation can also be produced by natural astrophysical processes – that's the reason the instruments exist in the first place. So the trick lies in looking for a mysterious pattern in the radiation which isn't consistent with any known natural origin – and then working out if it might have an artificial one.

That's exactly what happened with the discovery of so-called 'fast radio bursts', or FRBs, in 2007. These really are fast, lasting just a few milliseconds, but during that time they represent a huge release of energy. Very little of that energy reaches us, because it originates at such a great distance that it's diffused over an enormous volume of space. As far as we can tell, the sources of the FRBs aren't even in our own galaxy, but in other galaxies millions of light years away. There's no agreement on what causes them, but no shortage of speculations either. According to one of these, FRBs are 'spectral signatures of advanced interstellar spacecraft' of the kind Zubrin talked about.

The suggestion was made by astronomers Manasvi Lingam and Avi Loeb of Harvard University, in a paper published in the *Astrophysical Journal* in 2017. Bearing in mind the vast amount of energy needed to accelerate a starship too close to the speed of light, they argued that it might be impossible to produce all that energy on board the spacecraft itself. Instead, the energy would be generated on the planet of origin, pushing on the spacecraft with a directed blast of pressure – not of material particles, but a laser-like beam of radiation. The same principle, on a much more modest scale, has been considered for spacecraft propulsion inside the Solar System – using 'light sails' propelled by radiation pressure from the Sun.

While we may be able to build such a light sail in the near future, extending that to an interstellar spacecraft of maybe a million tons is way beyond our current capabilities. That doesn't stop us calculating what it would involve, though. In their paper, Lingam and Loeb do exactly that. They conclude that the key parameters of such a system, such as power and frequency, are a good match to the observed properties of FRBs. The fact that we just see them as millisecond-duration flashes doesn't mean they only operate for that brief a time, but that the beam is so tightly focused we only glimpse it for the fleeting moment that it lines up with the Earth.

Lingam and Loeb's argument doesn't prove that FRBs are the motive power source for alien starships – but it does make it a possibility. As Loeb was quoted as saying in the press release accompanying the paper:

> We haven't identified a possible natural source with any confidence. An artificial origin is worth contemplating and checking.

As true as that is, it remains just one of a number of tentative explanations for FRBs that have been put forward by different groups of scientists. Unsurprisingly, however, it's the one the mass media has been most enthusiastic about. As the Snopes fact-checking website pointed out in January 2019, press coverage of the subject has been heavily slanted towards 'the impression that a leading hypothesis for the source of these radio bursts was some sort of alien technology'. They're not just talking about tabloids either. Snopes even quotes a headline from the usually reputable *Guardian* newspaper: 'Mysterious Fast Radio Bursts from Deep Space Could Be Aliens'.

'Of course, anything *could be* aliens', as the Snopes reporter sardonically points out. And that brings us neatly on to the next topic ...

Ancient Aliens (seriously)

The idea that anything might be – and probably is – the work of aliens has been widely lampooned in an internet meme featuring Giorgio Tsoukalos, the wild-haired host of TV's *Ancient Aliens*. That series is based on the premise – already familiar from bestselling books like Erich von Däniken's *Chariots of the Gods?* – that extraterrestrials have been visiting Earth throughout its history, and actively helping to shape that history. Most famously, they're credited as the brains behind such wonders of the ancient world as the Egyptian pyramids. Most people can see the fallacy here ('So you're saying the Egyptians were too stupid to build them on their own?'; 'Why did the aliens use a low-tech material like stone?'; 'Where are all the high-tech alien relics?') – hence the popularity of the meme.*

As a consequence, *Ancient Aliens* tends to have the opposite of its intended effect. Instead of persuading people the Earth has been visited by aliens in the past, they go away convinced that it's a ridiculous notion. But it isn't. The ridiculous thing is the way the show connects that single obsessive idea to a whole range of myths, legends, structures and artefacts that already have perfectly adequate

---

* Perhaps the most pertinent version has Tsoukalos saying 'I don't know – therefore aliens'.

explanations in terms of the cultures they originated in. Shorn of all that silliness, however, there's nothing wrong with the idea that aliens from somewhere else in the galaxy might have visited the Solar System at some point in its 4.5 billion-year history.

If they did, it's most likely they didn't come in person, but sent robot probes instead. After all, that's the way we explore the Solar System today – and it makes a lot of sense. Machines require far less energy and resources to sustain them over a long voyage, and it's reasonable to suppose that with further improvements in technology they could be made extremely small and lightweight.

Although they'd still be limited to travelling at less than the speed of light, a fleet of robot probes would be capable of exploring the galaxy in a surprisingly short time. The key trick is to make them self-replicating. When a probe reaches a suitable planetary system, it uses the materials there to make further copies of itself, which then go off in different directions. Robot spacecraft of this type are referred to as von Neumann probes, after the physicist John von Neumann, who first suggested the idea in the 1960s.

Von Neumann's original description of how such probes might operate was highly idealised, because – like many physicists – he was more interested in principles than in practicality. The task of working out if and how such a probe might actually be built had to wait until an engineer, Robert Freitas, decided to tackle it. His report on 'A Self-Reproducing Interstellar Probe' was published in the *Journal of the British Interplanetary Society* in 1980. It's still only a paper study, but it does go into detail on things like dimensions, energy requirements and raw materials – and how they might be found and extracted in a new planetary system.

The result is, I suspect, far more complex than von Neumann imagined. The probe needs a large number of subsystems, which can be divided into four broad categories. First there is propulsion, just to get from one planetary system to another, and to decelerate and land at its destination. Then there's the scientific payload: the various sensors and instruments needed to explore the new system when it gets there. Just as important is a communication system, to relay the findings back along the chain of probes to the world of origin. And finally, there's all the mining and factory equipment needed to fabricate the necessary number of duplicate probes.

In spite of all this complexity, Freitas concluded that such a probe was a viable proposition – and he was thinking in terms of what was potentially doable in 1980, not extrapolating to hypothetical future technologies. One of the most surprising things about his calculation is the enormous mass of the resulting spacecraft. He estimated this at 10 million tonnes – 20 times the size of a supertanker – though that figure might come down if the builders had access to more advanced technologies. Another surprise is the length of time it would take to fabricate a duplicate probe – maybe 1,000 years, although that's still a lot less than the time it would take the probe to travel from one planetary system to another.

So how might the process work? The following table shows a highly simplified example. Imagine a hundred probes are sent out, travelling at, say, a tenth the speed of light. After a hundred thousand years, assume that one in ten of them has found an interesting planetary system, and used the resources there to make a hundred further copies of itself. If the same pattern is repeated over and over, it might be possible to explore a billion planetary systems in less than a million years – the blink of an eye in cosmic terms.

Exploring the galaxy with self-replicating von Neumann probes

Time (in years)	Number of planetary systems explored	Number of probes sent out
0	1	100
100,000	10	1,000
200,000	100	10,000
300,000	1,000	100,000
400,000	10,000	1,000,000
500,000	100,000	10,000,000
600,000	1,000,000	100,000,000
700,000	10,000,000	1,000,000,000
800,000	100,000,000	10,000,000,000
900,000	1,000,000,000	100,000,000,000

Based on this reasoning, it's not stretching things too much to suppose that one or more alien spacecraft have visited the Solar System at some point in the past. Having completed its mission, it might even still be here. The idea of a long-dead alien spacecraft lurking somewhere among the planets sounds like science fiction – and indeed it did crop up in a 1953 short story by Arthur C. Clarke called 'Jupiter Five'. The title comes from an earlier designation for one of the moons of Jupiter, now called Amalthea. It's much smaller than the other moons known at the time (Clarke quotes a diameter of 30 km, although it's actually quite a bit bigger than that), and the story has a group of astronauts discovering it to be an ancient, abandoned spaceship.

A few years later, the Russian astronomer I.S. Shklovskii made a similar suggestion concerning the small Martian moon Phobos – and this time it was a serious scientific hypothesis, not a sci-fi story. Shklovskii was struck not just by the small size of Phobos, but also by its low density and apparent peculiarities in its orbital motion. He deduced – incorrectly, we now know – that this miniature moon was hollow, and therefore almost certainly artificial. It was a suggestion that was taken quite seriously for many years, until better imaging revealed Phobos to be an irregularly shaped, asteroid-like chunk of rock – something that is also true of Amalthea, for that matter.

In reality, von Neumann probes are likely to be much smaller than Phobos or Amalthea. Even the 10 million tonnes estimated by Freitas is only a millionth of the mass of Phobos. Combined with the fact that, if they arrived millions of years ago, they are probably totally inert now, this makes searching for them next to impossible. The situation is reminiscent of an amusing scenario once put forward by the philosopher Bertrand Russell:

If I were to suggest that between the Earth and Mars there is a china teapot revolving about the Sun in an elliptical orbit, nobody would be able to disprove my assertion provided I were careful to add that the teapot is too small to be revealed even by our most powerful telescopes.

Russell's aim was to show how certain types of assertion can never be proved false. He was thinking in purely philosophical terms, but the particular example he chose has a practical significance for us. It highlights the fact that we can never be sure there isn't a small artificial object lurking somewhere in the depths of the Solar System.

The fact is, if we ever detect such an object it will almost certainly be by pure chance, rather than the result of a methodical search. In an earlier book in this series, *Cosmic Impact*, I described the efforts under way to detect and track any objects large enough to cause a global disaster if they crashed into Earth. But an alien space probe is likely to be tens or hundreds of times smaller than that. It will be centuries before we have the technology to map all the objects of that size in the Solar System.

While self-replicating von Neumann probes are likely to be small, that's not necessarily true of alien spacecraft in general. With something large enough to contain a crew of human-sized occupants, current methods of detecting space hazards might be good enough. It's just remotely conceivable, in fact, that they've already been successful.

In October 2017, an object called 'Oumuamua hit the headlines when it passed through the Solar System on a course that originated in – and headed back out into – interstellar space. On top of its unusual trajectory, 'Oumuamua had a strikingly elongated shape that cried out 'spaceship' to many people. That's even tentatively hinted at in the name it was given, which means 'advance scout' in Hawaiian.

'Oumuamua came and went so quickly that astronomers didn't get a good look at it. All the evidence they did manage to acquire, however, was consistent with a rocky asteroid-like body. No radio signals or excess infrared radiation were detected, and its trajectory was almost exactly what would be expected of an unpowered object coasting along in the Sun's gravitational field. There was a very slight additional acceleration, but this could be accounted for by 'outgassing' of volatile chemicals from the rock – an entirely natural effect.

But what if 'Oumuamua didn't have any volatile chemicals? The following year, an alternative explanation for the excess acceleration surfaced. A paper published in the *Astrophysical Journal* in November 2018 proposed that it could have been the result of the Sun's radiation pressure pushing on 'Oumuamua. That's essentially the same physical mechanism as the 'light sail' theory Lingam and Loeb put forward to explain FRBs, as mentioned earlier in this chapter. While the effect in 'Oumuamua's case might be perfectly natural, it might be artificial – as the paper's authors point out:

> Considering an artificial origin, one possibility is that 'Oumuamua is a light-sail, floating in interstellar space as debris from an advanced technological equipment … A more exotic scenario is that 'Oumuamua may be a fully operational probe sent intentionally to Earth vicinity by an alien civilisation.

And who are those authors, exactly? Their names are Shmuel Bialy and Avi Loeb – and yes, that's the same Avi Loeb who co-authored the FRB light sail paper. I'll confess at this point that the discussion of alien city lights and industrial pollution at the start of this chapter was also drawn from work by Loeb and his team. So you may get the impression that he's the mainstream science community's answer to Giorgio Tsoukalos, shouting 'It's aliens!' at every hint of an unexplained mystery.

That's not really fair, though. Loeb is employed by one of America's most prestigious universities, Harvard. He's the author, or co-author, of hundreds of papers on astrophysics – many of them speculative or provocative in nature, but only a small fraction of them touching on aliens or extraterrestrial technology. Even the ones that do – such as the FRB and 'Oumuamua papers – don't actually say 'Look, aliens', but 'Look, aliens are one possible explanation that can't be ruled out at this stage'. That, in a nutshell, is the difference between science and pseudoscience.

So Where Are They?

Seventy years after Fermi first voiced his 'paradox', the question he asked hasn't gone away. Where are the aliens? Radio telescopes are far more sensitive today than they were then, but they still haven't detected any obvious messages from extraterrestrials. There have been tantalising hints of things that might be alien technology – from FRBs to Tabby's Star to 'Oumuamua – but they all have alternative, more mundane explanations. So it's time to revisit Fermi's paradox and look at some of its possible resolutions.

Actually the hard work has already been done for us, by the philosopher of science Milan Ćirković, in his excellent book *The Great Silence* (2018). Ćirković not only delves into the philosophical nuances of Fermi's paradox far more deeply than we need to here, but he goes on to provide a critical survey of the numerous ways out of it that have been proposed over the years. He groups these into four broad categories: anti-realist solutions, anti-Copernican solutions, catastrophic solutions and logistic solutions. Without digging too far down into the weeds, let's take a brief look at how each of those works.

Although UFO believers would get hot under the collar if you said it to their faces, their worldview is the archetypal 'anti-realist solution'. If they're correct in everything they believe, then reality simply doesn't work the way that scientists and philosophers think it does. The only way to reconcile the mass of conflicting witness accounts concerning the appearance and behaviour of UFOs, and their occupants, is to assume the aliens have a quasi-godlike power to manipulate human perceptions and memories – not to mention photographic and documentary evidence.

That doesn't sound likely – but just like Russell's teapot, it's impossible to disprove. In a similar vein, and rather more acceptable to serious philosophers like Ćirković, is the simulation hypothesis. Here is his summary of it:

Physical reality we observe is, in fact, a simulation created by Programmers of an underlying, true reality and run on the advanced computers of that underlying reality … We cannot ever hope to establish the simulated nature of our world, provided that the Programmers do not reveal their presence.

That's reminiscent of the *Matrix* movies – or, for that matter, the Chinese legend of the philosopher Chuang Tzu, who could never decide if he was a human being dreaming he was a butterfly, or a butterfly dreaming he was a human being. The problem is, the moment we suspect our senses aren't giving us an honest picture of reality, anything becomes possible. Science can't deal with that – so we'll just have to hope it isn't the case, and move quickly on.

Ćirković's second option is that the basic Copernican principle – that there's nothing special about human beings or the planet Earth – is wrong. Maybe we really are special. This doesn't necessarily mean – although it might do – that we're somehow the 'chosen ones' of the universe, as people tended to believe in medieval times. A more modern approach would simply say that intelligent civilisation is highly unlikely – the result of a whole series of improbable events and coincidences – and we're just lucky that it all came together in the right way on Earth. By this argument, the dice would be unlikely to fall in the same way elsewhere in the galaxy. It's an idea that quite a few people, including Ćirković himself, take very seriously.

It's called the 'rare Earth hypothesis'. In their book *What Does a Martian Look Like?*, Jack Cohen and Ian Stewart say the following about it:

> It makes you think we only made it by the skin of our teeth. All those obstacles to overcome. All those ways for incipient life to fail. How likely is it that life could gain a toehold anywhere else? Earth is perfect, ideal for life. It may well be the only Earthlike planet in the universe.

They then devote the next dozen pages to an exhaustive, and very witty, demolition of the rare Earth hypothesis. Basically they pick out the central flaw in the argument, and keep hammering away at it until there's nothing left. The flaw is that the incredibly improbable chain of events that led to human civilisation doesn't mean there aren't millions of other chains of events that might lead to a functionally equivalent – but biologically different – civilisation. It's the whole chain that's different, and you can't expect to get from one to the other simply by focusing on one link at a time – which is what the 'rare Earth' proponents tend to do. Elsewhere in their book, Cohen and Stewart liken the problem to trying to convert a Toyota car into a Jaguar:

> If you try to replace the crankshaft of a Toyota with one from a Jaguar, it won't work. But that's because you do not get from a Toyota to a Jaguar by changing one tiny thing. … You can get from a Toyota to a Jaguar, but only by making a coordinated suite of changes.

Personally I agree with Cohen and Stewart on this, and I find the rare Earth hypothesis much harder to swallow than, say, the simulation hypothesis. I just don't like the idea of uniqueness. If you look at everything else we can see in the universe, there's no other type of object we only see one example of. Yes, specific examples always differ from each other in detail, but there's nothing as unique as the rare Earth hypothesis says the Earth is.

Having said that, it was probably a red herring to bring up the Copernican principle in this context anyway. The

Copernican principle says there should be other civilisations like ours – but that's not what we're looking for. There might be another such civilisation just ten light years away and we'd never know – because it wouldn't produce any readily detectable technosignatures at that range. What we're really talking about is an *advanced* civilisation – far more advanced than our own, such as might exist in our own far future. So the real issue is how likely such a civilisation is, and how long it's going to last.

This brings us to the next item on Ćirković's list – 'catastrophic solutions' to the Fermi paradox. These cover a wide range, some more obvious than others. At the obvious end, there are any number of planetary-scale disasters that could send life on Earth back to square one, from natural ones – such as a cosmic impact of the type that wiped out the dinosaurs – to self-inflicted ones, like nuclear war or anthropogenic climate change. But there are less obvious possibilities that are only 'catastrophic' in the sense that they'd put an end to our technological expansion.

Of the many great civilisations that have existed on Earth over the last few thousand years, ours is the only one to have developed the kind of technology, and the scientific worldview, that's relevant to space exploration. Who's to say future civilisations will follow in the same mould? Even today, there are numerous non-Western cultures that put other factors – spirituality in particular – above technological attainment. So maybe that's the shape of things to come. If alien civilisations generally come to the same conclusion, it would provide yet another resolution of Fermi's paradox.

We still haven't exhausted all the possible solutions. The final category is the one Ćirković calls 'logistic solutions'. According to this view, the reason we don't see advanced civilisations is because they've all migrated to a different part of the galaxy. Maybe they've headed to the galactic centre, where stars are packed more densely and material resources are more plentiful. Or perhaps they congregate around black holes, where the inflow of matter provides the most efficient source of energy in the known universe.

The fact is, we just don't know what a sufficiently advanced civilisation would see as a desirable neighbourhood. They may be attracted to regions of current star formation, or the site of recent supernova explosions, for reasons we can't begin to fathom. But these logistic arguments always boil down to the same thing: that we're living in an unfashionable galactic backwater, where no advanced alien would be seen dead.

I may have given the game away when I said that Milan Ćirković is a philosopher – because once the philosophers have got hold of a question, you know there's little chance it will ever be answered to everyone's satisfaction. There's no universally agreed explanation for Fermi's paradox, and probably never will be. But its basic assumption remains as true as ever: there's no unambiguous evidence for the existence of advanced civilisations elsewhere in the galaxy. So perhaps it's time to lower our sights a little, and consider the possibility of much more lowly forms of extraterrestrial life.

STARTING SMALL 5

After all the excitement of interstellar messages and alien megastructures in the previous two chapters, it may seem like a backward step to start talking about far more primitive forms of life. In a sense it is, but the emergence of such forms on Earth was the first step on the only path we know of that leads to intelligence.

Actually there's sound logic behind the ordering of the chapters. Virtually all the science in this chapter and the next is newer than anything we've talked about so far. Concepts like the Drake equation, Dyson spheres and von Neumann probes all date from the early 1960s, while the search for exoplanets and possible 'biosignatures' only took off in the 21st century. Equally new is the study of extremophiles here on Earth, which show us that the range of environments in which life can thrive is much wider than used to be imagined.

The discovery of biosignatures indicating the presence of life on an exoplanet – or maybe even on a planet or moon here in the Solar System – may not seem as exciting as picking up a radio message from aliens, or spotting an artificial megastructure around a nearby star. But the fact that primitive lifeforms have existed on Earth for so much longer than advanced ones implies that they'll be a much more common occurrence in the galaxy – and hence, in principle, easier for us to find. And the way we go about finding them has an excitement of its own – albeit the excitement of a detective story rather than that of a sci-fi blockbuster.

In the Beginning

If you look out of your window, you'll see plenty of evidence of life on Earth – people, dogs, cats, birds, insects, flowers, trees, grass – but that's only the tip of the iceberg. All those examples are complex multicellular organisms of a type known as eukaryotes. But they only represent a small fraction of the entire domain of eukaryotes, most of which are microscopically small – sometimes, as in the case of an amoeba, just a single cell. And it doesn't end with eukaryotes. There are two other domains of life, bacteria and archaea, collectively referred to as prokaryotes. These too are single-celled organisms, but much smaller and simpler than even the most primitive eukaryotes.

There's one thing that all forms of terrestrial life have in common, and that's DNA. This is an enormously complex carbon-based molecule – made up of billions of atoms – that's present in every living cell, and which determines the form of an organism through the genetic data it contains. Remarkably, the DNA molecule is capable of self-replication;

it can make copies of itself. Usually these copies are perfect, but occasionally mutations can occur – and if the mutations are favourable, these can ultimately lead to the evolution of a completely new species. By tracing the similarities and differences between the DNA of different species, it's clear they've all evolved from a single common ancestor – as shown in the below 'family tree' of life on Earth.

The most familiar names on the family tree can be found in the top right corner: 'animalia' (animals), 'plantae' (plants) and fungi. To our eyes, those three groups of large, multi-celled eukaryotes appear to dominate life on Earth – but that's not actually true. For the first 2 billion years after life started around 4 billion years ago, the world belonged to prokaryotes – there were no eukaryotes at all. Even after the latter made their appearance, it was another billion years before they evolved into non-microscopic forms. To this day,

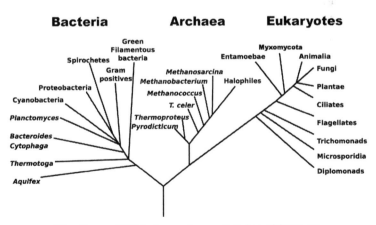

The 'tree of life' shows how all living things on Earth are descended from a common ancestor.

(NASA image)

prokaryotes vastly outnumber eukaryotes, even in environments where the latter thrive.

If that was as far as it went, it would be an interesting piece of trivia, but hardly worth mentioning in a book about astrobiology. What changes everything is the fact that some prokaryotes have carved a niche in environments that eukaryotes like ourselves would think of as totally inhospitable to life.

This is a relatively recent discovery – the word 'extremophile', which describes such organisms, was only coined in the 1970s – and in some ways it's as revelatory as the discovery of life on another planet would be. We now know there is virtually nowhere on Earth that life, at the simple level of archaea and bacteria, hasn't found a way to thrive. We're talking about incredibly alien environments here – from our point of view, that is, since the extremophiles would feel the same way about where we live.

'Extremophile' means loving extremes – which may be extremes of acidity or alkalinity, high or low temperatures, the complete absence of light or the presence of intense radiation. One extremophile, *Deinococcus radiodurans*, can live happily inside a nuclear reactor. Other species are found in salt lakes, in rocks deep within the Earth's crust, or congregating round hydrothermal vents on the ocean floor, where the pressure is crushingly high and the water is superheated to hundreds of degrees Celsius.

This last environment is particularly significant, because one of the leading hypotheses for the origin of life on Earth is that it started in the vicinity of exactly this kind of vent. It turns out that some of the critical biochemistry that goes on inside living cells might also take place naturally around a hydrothermal vent, without the need for a cell per se. If this is the case, it provides a neat explanation of another mystery – why the cell structures of bacteria and archaea are so different, despite the fact that their DNA shows them to have a common ancestor. Perhaps that ancestor didn't have cells at all, because it got all the biochemistry it needed from a hydrothermal vent.

But just what is 'biochemistry' anyway? We know what chemistry is, and we know what biology is, but at what point does chemistry turn into biology? In one sense, DNA is just another chemical – but in the context of a living cell it's self-replicating, and capable of adapting over successive generations to thrive more successfully in its environment. In other words, it meets NASA's definition of life – 'a self-sustaining chemical system capable of Darwinian evolution' – which we met in the first chapter.

So how easy is that crucial step from inanimate chemistry to a living organism? To some extent that's a question for the philosophers, and one answer they've come up with is the 'continuity thesis' – described in Milan Ćirković's book *The Great Silence* as 'the assumption that there is no unbridgeable gap between inorganic matter and living systems, and that under suitable physical conditions the emergence of life is highly probable'.

As a physicist, I like this idea. From a physics perspective, everything that happens in the universe does so because it involves energy flowing in accordance with the laws of thermodynamics. Under these circumstances, there are no ifs or buts – when something is thermodynamically favourable, it happens. Perhaps life falls into the same category – albeit in a way that's somewhat different from usual.

The normal effect of thermodynamics is to bring different parts of a system into equilibrium – with no net energy flows – as quickly as possible. That's why your coffee gets cold if you put it down for a few minutes to answer the phone. But living things are different – they somehow manage to remain in a non-equilibrium state, with highly organised internal energy flows, for as long as they remain alive. Yet maybe the same laws of thermodynamics lie behind this behaviour too. That's the essence of a theory called 'dissipation-driven adaptation', originated by Jeremy England at the Massachusetts Institute of Technology. Here's how Brian Clegg described England's work in *BBC Focus* magazine:

> Living things are in a non-equilibrium state, taking energy from sources such as sunlight and food and pushing that energy out – 'dissipating' it – into their surroundings. This enables the living organism to ... grow and build structure. And it is the physics of such non-equilibrium states that England and his team investigate, by using computer simulations to look for situations where life-like behaviours emerge spontaneously.

England's theory hasn't quite made it into the textbooks yet, but the results of his simulations look highly promising. So it may well turn out that life appeared and spread on Earth for the simple reason that it was thermodynamically favourable. If that's the case, then we would expect something very similar to happen on any other planet where the same conditions prevail. The same point was made, rather more poetically, by Professor Brian Cox in the BBC documentary series *Wonders of Life* in 2013:

> Far from being some chance event ignited by a mystical spark, the emergence of life on Earth might have been an inevitable consequence of the laws of physics – and if that's true then a living cosmos might be the only way our cosmos can be.

Just a moment, though – isn't there an obvious flaw in the idea that it's easy to create life from inanimate matter? If it's so easy, why don't we see it happening on Earth today? As it happens, there's a good answer to this one. The conditions on Earth 4 billion years ago, when life first appeared, were significantly different from today. Most strikingly – to an organism that thinks of oxygen as an essential prerequisite for staying alive – there was no oxygen at all in the atmosphere of the early Earth.

The absence of readily available oxygen had two consequences. The first has to do with chemistry. To get the raw materials for life, we need to start with basic constituents like carbon and hydrogen and turn them into progressively more complex molecules. But you can't do that if there's plenty of oxygen around, because it jumps onto the hydrogen to make water (H_2O), and onto the carbon to make carbon dioxide (CO_2), and everything stops there. To build up complex molecules, you have to do it without the interference of oxygen.

Atmospheric oxygen has a second consequence, too. Both in its breathable form (O_2) and as ozone (O_3) in the upper atmosphere, it's very good at blocking out high-energy ultraviolet (UV) radiation from the Sun. That's good news for us – and for most other lifeforms on the surface of the planet* – because UV is highly damaging to DNA. On the other hand, it's not so helpful if you're trying to synthesise complex organic molecules, because that requires a constant input of energy – for which unfiltered UV radiation would be ideal.

All in all, the oxygen-free early Earth looks like a much more promising environment for the emergence of life than the Earth of today. It's an idea that was put to the test in 1952 by a PhD student named Stanley Miller, under the supervision of Harold Urey at the University of Chicago. Miller and Urey set up a table-top experiment to duplicate, at a basic level, the conditions that prevailed on the primordial Earth. Their apparatus consisted of a flask of water connected to a flask of gases – methane, ammonia and hydrogen – representing the early, pre-oxygenation atmosphere. The energy input, in that original Miller–Urey experiment, was provided by an electrical spark – meant to represent lightning flashes – although later reruns replaced this with a UV source, which is probably more realistic.

If the Miller–Urey experiment was a 1950s B-movie, instead of 1950s real life, it would have produced a living protoplasmic blob that went on to take over the whole of Chicago. Of course, that didn't happen – and it was never meant to. The aim of the exercise wasn't to create life, but the chemical building blocks of life. In this sense it was a

---

* Yes, you guessed it – some extremophile species actually like UV.

resounding success, producing key constituents of living cells called amino acids. The first of these was detected within two days, and a dozen more varieties by the end of the first week. Eventually the experiment yielded all 20 amino acids commonly found in life on Earth.

While the original Miller–Urey experiment produced all the amino acids necessary for life on Earth, there was no evidence of another essential component: the four nucleobases that make up the building blocks of DNA. It would be a long time before that shortcoming was remedied, but eventually it was. The DNA nucleobases were synthesised, for example, in a 2017 rerun of Miller and Urey's experiment which added a subtle new ingredient to the mix: 'powerful laser discharges to simulate the plasmas resulting from asteroid impact shock waves'.

That may sound far-fetched, because asteroid impacts are – thankfully – few and far between these days. But that's another thing that was different 4 billion years ago. The planets had only just finished forming, and the Solar System was still cluttered with huge chunks of left-over rock. As a consequence, the rate of impact events was much higher than it is today. It's not unreasonable to imagine that those impacts played a role in getting the first biochemical reactions going.

I said earlier that the one thing that all terrestrial organisms have in common is DNA: a specific type of complex molecule made up of countless combinations of four simple building blocks, or nucleobases. Is that going to be an essential requirement for life anywhere else in the galaxy? In a broad sense, the answer is probably yes. Any 'self-sustaining chemical system capable of Darwinian evolution' is going to need a chemical structure capable of encoding complex genetic information, and a natural mechanism for making copies of itself. In specific terms, however, there's nothing to say that chemical structure has to be identical to our local brand of DNA.

The study of non-DNA biochemistry is called xeno-biology, from the Greek *xenos* meaning 'stranger'. The word was originally coined by the science fiction author Robert A. Heinlein, in his 1954 novel *The Star Beast*. Heinlein used it to mean the study of extraterrestrial lifeforms – rather like real-world astrobiology, except that Heinlein's xenobiologists are space travellers who can get up close and personal with their subject matter, rather than having to search for it with telescopes. Today, however, xenobiology has broadened out in meaning to cover any kind of biology that's different from our own – and scientifically feasible – regardless of whether examples of it actually exist or not.

As it turns out, there are quite a few ways that life might work without using our own particular kind of DNA. One example, which popped up in the news as I was writing the first draft of this book, is 'hachimoji DNA'. That sounds very exotic, unless you happen to know a few words of Japanese. *Hachi* means 'eight' and *moji* means 'letter'. Traditionally the standard four DNA bases are known by letter codes – A, C, G and T – and the hachimoji variety simply adds another four: B, P S and Z. It's not just a theoretical concept, either. The NASA-funded team behind the discovery actually created hachimoji DNA in their laboratory. It's capable of storing and transmitting genetic information just like regular DNA, but it does it in a different way. And that's only the start; scientists believe there are many other ways to provide essentially the same functionality.

This sort of research has obvious implications for the way we look for extraterrestrial life. As the director of NASA's astrobiology programme, Mary Voytek, put it:

> Incorporating a broader understanding of what is possible in our instrument design and mission concepts will result in a more inclusive and, therefore, more effective search for life beyond Earth.

Coming back to our own planet, to a large extent it's been shaped by the presence of life. The first bacteria produced oxygen as a waste product, and that created an atmosphere – and a natural sun-screen – that was more hospitable for the development of increasingly advanced forms of life. But the change happened over a long period of time, and it's a mistake to think that a 'habitable world' has to look exactly like the Earth does today. If we widen our perspective – in both time and space – we can find conditions closely resembling the early Earth elsewhere in the Solar System.

Planetary Neighbours

Looking at all the known habitats of life on Earth, either now or in the past, three factors emerge as essential prerequisites for 'life as we know it'. The first is a supply of energy – fuel to power biological processes. We know where we get our energy from: food – because it says so on the label. For example, a 250 g steak provides around 1,200 kilojoules of energy. That energy in turn came from further down the food chain – the grass that cows graze on. But the grass got its energy from outside the terrestrial life cycle – direct from the Sun, via photosynthesis. In the same way, the ultimate source of energy for most lifeforms is heat and light from the Sun – although in some cases it can be geothermal energy originating inside the Earth itself.

The second prerequisite is a stock of the basic elements needed to build amino acids and nucleobases – carbon, hydrogen, oxygen and nitrogen. Those are all common elements throughout the Solar System, often occurring in the form of simple molecules like carbon dioxide (CO_2), methane (CH_4) and ammonia (NH_3) – although simple amino acids are probably also quite widespread, since they're often found in meteorites, and were detected in comet 67P when it was encountered by the Rosetta spacecraft in 2014.

So far, there's nothing to make Earth any different from the other planets. They all have the necessary raw materials, and they all receive their share of sunlight. The tricky part comes when we move onto the third prerequisite for life: liquid water. Perhaps less obvious than the first two, its crucial role is in providing a solvent in which the necessary biochemical reactions can take place. It's possible that other liquids might be able to play this role, but that's speculation and not at all certain. If we stick to what we know for a fact, then we're forced to assume that liquid water is essential for life.

Unfortunately, it's surprisingly difficult to find liquid water on the other planets of the Solar System. The obvious place to look would be Earth's nearest neighbours, Venus and Mars. They're both rocky planets of similar composition to the Earth, with Venus being about 5 per cent smaller than Earth, and Mars just over half its size. The key difference, however, is in a planet's distance from the Sun – and hence the amount of solar heating it receives. Mars gets less than the Earth, being 50 per cent further away from the Sun, while Venus – 25 per cent closer – gets more.

**Comparison of Venus, Earth and Mars,
pictured on the same scale.**
(NASA images)

The situation on Venus is exacerbated by the fact that it's enveloped by a dense atmosphere composed largely of carbon dioxide. This produces a 'greenhouse effect', trapping the Sun's heat rather than allowing it to dissipate back into space. The result is a surface temperature around 460°C – obviously far too hot for liquid water.

That hasn't always been the case, though. Billions of years ago, before the greenhouse effect took hold, Venus would have been much more Earth-like – and might have developed a similar form of simple, single-celled life. It's just conceivable that the descendants of that early life might still survive, not on the surface of Venus, but in its upper atmosphere. The temperature and pressure at an altitude of 50 km are similar to those at the surface of the Earth, and the dense clouds there contain all the basic elements needed to sustain life – and possibly tiny droplets of liquid water, too. So we can't completely cross Venus off our list of places to look for life in the Solar System – but it's hardly a frontrunner.

Mars is a different matter. It's a place where many scientists positively *expect* to find evidence of life – if not there today, then at some stage in the distant past. As with Venus, the climate on Mars was probably much more Earth-like around the time that terrestrial life first appeared, but since then it's gone in the opposite direction. Mars now has a very thin atmosphere, having lost most of it on account of its smaller size and weaker gravity. That thin atmosphere, coupled with the greater distance from the Sun, makes Mars a chilly place, with surface temperatures rarely rising above 0°C. The combination of low temperature and low atmospheric pressure mean there's virtually no chance of liquid water on its surface.

It was different in the past. The Martian landscape shows numerous signs of having been shaped by water, from valleys carved out by ancient rivers to dried-up lakebeds. But judging by the number of impact craters that overlie such features, they must have been formed a long time ago. According to the best current estimates, we need to go back 3 billion years in order to find large amounts of water on Mars. As long ago as that sounds, however, there had already been life on Earth – in the form of primitive prokaryotes – for a billion years by that time. So it's entirely possible that a similar situation existed on Mars.

It appeared that this very possibility had been confirmed in a dramatic incident that occurred on 7 August 1996. It was the closest the world has yet come to the sci-fi cliché where the President of the United States emerges onto the White House Lawn to announce the arrival of extraterrestrials. The President at the time was Bill Clinton – and he was talking not about humanoid aliens in a spaceship, but microscopic ones in a meteorite. Here is an excerpt from his speech:

> It is well worth contemplating how we reached this moment of discovery. More than 4 billion years ago this piece of rock was formed as a part of the original crust of Mars. After billions of years it broke from the surface and began a 16 million-year journey through space that would end here on Earth. It arrived in a meteor shower 13,000 years ago. And in 1984 an American scientist on an annual US government mission to search for meteors on Antarctica picked it up and took it to be studied. Appropriately, it was the first rock to be picked up that year – rock number 84001.
>
> Today, rock 84001 speaks to us across all those billions of years and millions of miles. It speaks of the possibility of life. If this discovery is confirmed, it will surely be one of the most stunning insights into our universe that science has ever uncovered. Its implications are as far-reaching and awe-inspiring as can be imagined. Even as it promises answers to some of our oldest questions, it poses still others even more fundamental.

Some of these facts were uncontroversial. It's perfectly possible for small rocks – this one was just under 2 kilograms – to be flung all the way from Mars to Earth by a powerful impact event. We know it came from Mars because of its distinctive composition, and the timeline of events recounted by Clinton was verified by radiometric dating. It's also uncontentious that, while the rock was sitting on Mars, it was fractured and permeated by flowing water sometime between 4 and 3.6 billion years ago.

The contentious part is the idea that those water-filled fractures were subsequently colonised by Martian bacteria. President Clinton's speech was prompted by an academic paper, published in the journal *Science* the previous day, that claimed just that. Called 'Search for Past Life on Mars: Possible Relic Biogenic Activity in Martian Meteorite ALH84001', the paper was the work of a NASA team led by David McKay. Their argument wasn't based on any single, clinching piece of evidence but on a whole string of suggestive facts.

To start with, ALH84001 contains not only amino acids – which isn't unusual for meteorites – but also other compounds that are typically found on Earth as the decay products of microorganisms. Another suggestive indicator was the mineral magnetite, which occurs both naturally and in bacteria on Earth – but the form in ALH84001 looked more like bacterial magnetite than the naturally occurring kind. Most strikingly of all, photographs taken with a scanning electron microscope revealed structures that looked for all the world like the fossils of tiny bacteria, around 20 to 100 nanometres* in size.

---

* A nanometre is a millionth of a millimetre.

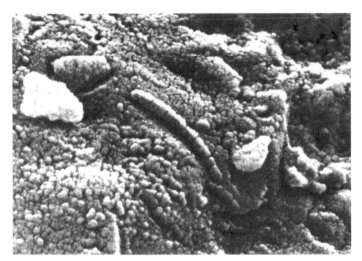

**Electron microscope image of meteorite ALH84001,
showing tiny structures that look like fossilised bacteria.**
(NASA image)

McKay and his co-workers concluded their paper with the following:

> None of these observations is in itself conclusive for the existence of past life. Although there are alternative explanations for each of these phenomena taken individually, when they are considered collectively, particularly in view of their spatial association, we conclude that they are evidence for primitive life on early Mars.

Not everyone agreed. To many scientists, the fact that there were 'alternative explanations' for all the observed evidence meant there was simply no need to resort to the idea they were produced by Martian bacteria. One of the strongest arguments against McKay's interpretation was the fact that no known life-forms on Earth are as small as 100 nanometres – although the smallest archaea and bacteria are only slightly larger than that.

As someone who is generally sceptical of sceptics – particularly knee-jerk debunkers who instantly try to throw a wet blanket on any provocative new idea – I'm amused by the fact that the ALH84001 sceptics fall into two mutually contradictory camps. One camp accepts that all the features identified by McKay's team originate on Mars, but argues that they have a non-biological origin. The other camp accepts the evidence for microorganisms in the meteorite, but claims that it's terrestrial contamination that occurred after it landed in Antarctica. They can't both be right – and in effect they're arguing against each other.

The fact is that McKay's hypothesis of fossilised Martian bacteria in ALH84001 remains a third possibility. Realistically, however, until or unless further evidence comes to light, it's less likely than the 'non-biological' or 'terrestrial contamination' hypotheses.

Suppose that life did evolve early in the history of Mars; what happened to it when the atmosphere leaked away and the surface dried up? It's easy to imagine that it all died out, but that's not necessarily the case. It's possible that Martian life carried on deep below the surface, where conditions are actually far more hospitable. If that idea sounds extraordinary to us, it's because, like most eukaryotes, we happen to live on the surface of a planet. But as we've seen, eukaryotes are in the minority – and an astonishing 70 per cent of all prokaryotes, both bacteria and archaea, actually live inside the Earth rather than on the surface.

Some of those subterranean organisms have found a home in deep caves, but others live in much smaller cracks and voids inside rock – more or less any space will do, as long as it contains water. All the other chemicals needed for life are there, while the necessary energy comes not from the Sun, but from the heat content of the rocks themselves. It's possible to imagine a similar 'deep biosphere' existing on Mars too – although it's not an easy thing to prove. Finding positive evidence would involve digging deep down into the Martian rock or venturing into its cave complexes – neither of which is likely to be feasible, even with robot explorers, for many decades.

It may come as a surprise, but of the half-dozen space-craft that have successfully landed on Mars in the last 30 years, not a single one of them was designed to search for signs of life. With a little luck the situation will change soon, with not one but two new rovers – one from NASA and one from the European Space Agency (ESA) – due to touch down on the Red Planet early in 2021. Both rovers will carry instruments capable of detecting subsurface water, as well as a range of 'biosignature' chemicals that might indicate the presence of hidden microorganisms.

That's still in the future, though. At the time of writing, if you want to find an experiment designed to search for life on Mars you have to go all the way back to 1976 – and NASA's very first Mars landers, Viking 1 and 2. They each carried a scoop to pick up a sample of Martian soil, which was then fed into a battery of scientific instruments. Three of these instruments were specifically designed to look for living organisms, by adding a mix of chemicals to the soil and then watching for a particular reaction deemed to constitute a 'positive' result.

The outcome was the same for both landers, in completely different locations on the planet. One experiment produced a negative result, the second was marginal – and the third was resoundingly positive. The successful experiment, called 'labelled release', involved feeding the sample with a nutrient mix containing a radioactive version of carbon,* and then testing for any gas emissions containing those radioactive carbon atoms. Both landers registered a huge amount of such gas – but was it really being 'exhaled' by Martian bacteria?

That was the original assumption made by the Viking science team, whose most famous member was Carl Sagan. Four years after the Viking landings, in his 1980 book *Cosmos*, he admitted the assumption may have been incorrect:

> Enormous efforts were made to build the Viking microbiology experiments and test them with a variety of microbes. Very little effort was made to calibrate the experiments with plausible inorganic Martian surface materials.

In other words, they hadn't given enough thought to the possibility of false positives. Because the other experiments were essentially negative, NASA never made a big thing about the labelled release result, and the general consensus now is that it was indeed a false positive. However, the exact mechanism that produced it has never been pinned down,

---

* Yes, that sounds a little cruel to me, too. It would probably be considered unethical today.

so the outside possibility remains that it really was caused by microbes in the soil.

For all its frustrating ambiguity, the Viking labelled release experiment is an excellent example of real science in action. The team started with the hypothesis that microbial life on Mars functions in a similar way to its terrestrial counterpart, and designed an experiment to detect one of the consequences of that hypothesis – that adding a nutrient solution to Martian soil should produce waste gases. When the experiment yielded a positive result – but not quite in the way that was expected – they looked for an alternative hypothesis that would fit the observed results better.

That's in total contrast to another, thoroughly unscientific, outcome of the Viking mission. Among the many photographs of the Martian surface taken by the Viking orbiters, one showed what – to human eyes – looks rather like a human face. It's really just a natural rock formation with a passing resemblance to a face (a resemblance that largely disappears in more recent, better quality images), but to some people it really was a giant face, which aliens carved into the Martian landscape in the distant past.

One of the most persistent champions of this idea was Richard Hoagland, who described in his 1987 book *The Monuments of Mars* how he had found numerous other artificial structures in the Viking photographs. As entertaining as these claims are, they wouldn't normally warrant a mention in a serious book on astrobiology. But Hoagland made another, completely different claim about alien life in the Solar System – and it's a much more interesting one. At the time he made it, most scientists found it just as outrageous as the Face on

Mars – but now they've changed their minds. To see why, we're going to have to travel half a billion miles beyond Mars.

Ice Worlds

There's very little water on the inner planets of the Solar System. That's obvious in the case of Venus and Mars – and Mercury, even closer to the Sun – but it's true of Earth too. Despite all those oceans, lakes, rivers and rain clouds, water accounts for only about a fiftieth of one per cent of the total mass of our planet.

The picture changes completely when you go further out, past the orbit of Mars. The outer Solar System contains enormous quantities of water. Jupiter's largest moon, Ganymede, and Saturn's largest, Titan – both half as big again as our own Moon – are close to 50 per cent water. But that's frozen water – ice – not the liquid water needed to sustain life. At five times Earth's distance from the Sun, the surface temperature of Jupiter's moons is something like –150°C, while Saturn's, twice as far away again, are 30 degrees colder still.

On paper, the moons of Jupiter and Saturn are well outside the Solar System's 'habitable zone' – the region where water can exist as a liquid, and hence life might flourish. If the Sun's radiation were the only source of heat, this would indeed be true. But there's another, far less obvious heat source – and that's gravity.

The pull of gravity causes a moon to orbit around a planet, rather than flying off into space. It's not a uniform pull over the whole moon, though. It's stronger on the side facing towards the planet than on the side facing away. There can be other fluctuations too, caused by the gravitational attraction of other moons orbiting the same planet. These variations combine to produce tidal stresses on the moon, analogous to the tides on Earth caused by the pull of our own Moon. But the tidal forces on the moons of Jupiter and Saturn are much greater – and they turn out to be a very effective source of heat in the form of volcanic activity.

This effect can be seen most dramatically in the case of Jupiter's innermost moon Io. It's the most volcanically active body in the Solar System, even more so than the Earth. In the vicinity of its volcanoes, the temperature can rise as high as 1,600°C. Liquid lava flows are constantly reshaping the surface, and the heat has – uncharacteristically for this part of the Solar System – boiled off any water Io ever had.

That's not the case with the next moon out, Europa. We've already met this one, on the very first page of this book, so you know it's going to be interesting. Personally, the first I ever heard of it was when I read Arthur C. Clarke's sequel to *2001: A Space Odyssey*, which he called *2010: Odyssey Two*. In my whole life, it's the only novel I've been so keen to read that I splurged out on the hardback the moment it came out – and the £7.95 it cost was a lot of money back in 1982 (but worth it). The novel makes the extraordinary assumption that Europa's icy surface conceals a huge ocean of liquid water, in which life thrives. That wasn't Clarke's own idea, though, as he explains in the book's afterword:

> The fascinating idea that there might be life on Europa,
> beneath ice-covered oceans kept liquid by the same Jovian

tidal forces that heat Io, was first proposed by Richard C. Hoagland in the magazine *Star and Sky* ('The Europa Enigma', January 1980).

Yes, that's the same Richard Hoagland who went on to write *The Monuments of Mars*. But unlike the Face on Mars, this particular off-the-wall notion of Hoagland's turned out to be a good one – a really good one. It's now widely accepted that Europa's icy surface does indeed conceal a huge subsurface ocean, which may contain twice as much liquid water as all the Earth's oceans put together. Coupled with the energy flows created by tidal heating, and the essential elements contained in the underlying rock, that means Europa has all the ingredients for life to develop.

Europa is such a promising habitat for life that several space missions to it are being planned. First, there's NASA's Europa Clipper. If this goes ahead – and that's always a big 'if' with NASA's more ambitious space programmes – it will launch in the mid-2020s. After a trip that may take as long as six years, it will probe Europa with an array of scientific instruments over the course of repeated flybys of the moon. As NASA scientist Robert Pappalardo explained in 2017:

> We're really trying to get at Europa's potential habitability, the ingredients for life: water, and whether there's chemical energy for life. We do that by trying to understand the ocean and the ice shell, the composition and the geology.

Around the same time that NASA plans to launch Europa Clipper, its European counterpart ESA is aiming to launch a spacecraft with the rather dubious name of JUICE (it stands for 'Jupiter Icy Moons Explorer'). This too would take a close look at Europa, although its ultimate destination is the next moon out, Ganymede, which is also believed to have a large subsurface ocean.

Most ambitious of all, NASA also has tentative plans for a Europa Lander, which would actually touch down on the surface in order to conduct in-situ experiments. This one, however, is even less certain to get off the ground than JUICE or Europa Clipper – and not just for the usual financial reasons. There's also the question of interplanetary ethics. If there's life on Europa, what right do we have to 'invade' its territory? It wouldn't just be a robot invasion, either. No matter how much a spacecraft is sterilised before launch, there will still be millions of terrestrial microbes happily clinging to it.

We have no idea what would happen if those microbes found their way into Europa's ocean. If it already had a thriving ecosystem, that might be destroyed or disrupted by the Earthly invaders – just as invasive species can cause huge damage to ecosystems on Earth. On the other hand, if there were no prior lifeforms on Europa, the Earthly bacteria might settle and thrive there – confusing the picture for future generations of scientists. These are eventualities that NASA takes seriously, and it's already taken steps to avoid them. When its Jupiter orbiter Galileo came to the end of its mission in 2003, it was deliberately destroyed – by crashing it into Jupiter – to ensure it could never end up contaminating Europa.

Fifteen years after Galileo's demise, a group of scientists reanalysing some of its data discovered a remarkable thing. At one point the spacecraft had flown straight through a plume of water blasted up from Europa – rather like an enormous version of a terrestrial geyser. Presumably this originated in the liquid water ocean deep inside the moon. If so, that's great news for future space missions, because it means they can study Europa's internal composition from space, without needing to land and drill down to the sub-surface ocean.

A similar geyser-like phenomenon can also be seen, even more spectacularly, on a moon of Saturn called Enceladus. It's much smaller than Europa – at 500 km, just a sixth of its diameter – but quite similar in appearance, with a smooth, ice-covered surface. A cluster of geysers near its south pole was discovered by NASA's Cassini spacecraft in 2005, and photographed on numerous subsequent occasions.

Enceladus's geysers are powerful enough to shoot mater-ial 900 km out into space – almost twice the moon's own diameter. Chemical analysis indicates the plume is 99 per cent water, along with several other ingredients. These include compounds of carbon, nitrogen, hydrogen and oxy-gen – so Enceladus has all the elements necessary to support terrestrial-type life. As with Europa, the most likely origin of the plume is a subsurface ocean, as shown schematically in the picture opposite. All in all – and despite its small size – Enceladus looks like another promising place to search for extraterrestrial life.

Before we leave the subject of ice worlds, there's just one more that's worth mentioning – and that's Saturn's largest moon, Titan. It's the only moon in the Solar System with a significant atmosphere – in fact the surface pressure is about

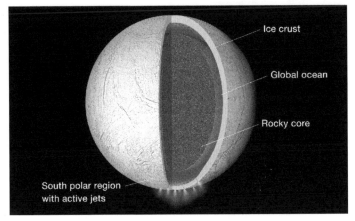

Assumed internal structure of Saturn's moon Enceladus (not to scale), showing postulated global subsurface ocean and geysers at the south pole.

(NASA image)

50 per cent higher than on Earth. And as on Earth, the commonest constituent of Titan's atmosphere is nitrogen, with the rest consisting largely of methane and other hydrocarbons. The annoying thing about the latter – from the point of view of an outside observer – is that they create dense, smoggy clouds which perpetually hide the surface from view.

The first glimpse of Titan's landscape had to wait until the Huygens lander parachuted down to the surface in January 2005, taking a number of panoramic photographs as it descended. Designed by ESA, Huygens hitched a ride to Titan on board NASA's Cassini spacecraft. After deploying the lander, Cassini went into orbit around Titan and took a series of high-resolution images of the surface – using radar, rather than visible light, to peer down through the cloud cover.

At first glance, the terrain revealed in the Huygens and Cassini photographs looks surprisingly Earth-like – but that impression is misleading. What look like rocky mountains are really made from deep-frozen ice, and what look like lakes and rivers are made from liquid methane. Titan has methane rain, too, falling from methane clouds.

From an astrobiological point of view, it's a frustrating situation. Titan's landscape looks complex enough that you feel it *ought* to have life. It's the only place in the Solar System, other than here on Earth, that has large bodies of liquid. The problem is it's methane – a gas under terrestrial conditions – rather than water. Titan even has a methane cycle, of evaporation and precipitation, exactly analogous to the water cycle on Earth. But the stark fact remains: without liquid water, the surface of Titan can't be home to life as we know it.

Of course, there's always the possibility that Titan has a subsurface ocean of liquid water, like its much smaller sibling Enceladus. If so, there's a chance that Earth-like life might be lurking there. A more way-out possibility is that a completely different form of biology has evolved on Titan's surface, using liquid methane as a solvent instead of water. There's even a tiny hint that this might indeed be possible.

In 2009, a PhD student at the University of Arizona, Sarah Hörst, ran a variant of the Miller–Urey experiment tailored to conditions on Titan. Most of the ingredients were the same – but one supposedly key one was missing. There was no liquid water in Hörst's version of the experiment. Yet it was successful – more successful, in a way, than Miller and Urey's original setup. As well as creating amino acids,

Hörst's experiment produced all four of the DNA nucleo-bases. It was the first time such chemicals had ever been made in the absence of liquid water – by scientists on Earth, anyway. But who's to say Titan didn't get there first?

So it's just conceivable that some kind of life has carved a niche for itself in the liquid methane lakes on Titan. But the fact remains that any organisms there – or on Enceladus or Europa or Mars – will almost certainly be microscopic in scale, on the same level as terrestrial archaea and bacteria. As interesting as that would be for scientists, it's not what most people mean when they get excited about extraterrestrial life. If we want to find more complex organisms, we're going to have to expand our horizons beyond the Solar System.

EXOPLANETS

6

As we saw in Chapter 2, the idea of planets orbiting other stars goes back several centuries. The earliest arguments were essentially philosophical, based on the Copernican principle that there's nothing unique about our own little corner of the universe. This view had thoroughly established itself by the time Bernard de Fontenelle wrote his book on the 'plurality of worlds' in the late 17th century (see page 20) – to the extent that the aristocratic lady his hero is lecturing beats him to the punch:

> 'Do I mistake,' cried the Marchioness, 'or do I see your drift? Are you not going to say the fixed stars are all suns … ? Our Sun enlightens planets; why should not every fixed star likewise enlighten planets?'

As time went on, such philosophical arguments were reinforced by other lines of reasoning, based on solid science. The 18th century saw the first serious attempt to explain the origin of the Solar System, which visualised the Sun condensing out of a primordial cloud of gas and dust called a nebula. The left-over bits of this, initially swirling around the Sun in a protoplanetary disc, eventually clumped together to form the planets. This is still considered a pretty accurate picture – not just for the formation of our own Sun, but of stars in general. And the idea that there will always be 'left-over bits' to form a planetary system isn't as glib as it sounds. It's a matter of basic physics.

One of the most important quantities in physics goes by the clunky-sounding name of angular momentum. It's essentially a measure of the amount of spin that an object has – and like energy, it's rigorously conserved. Whatever spin the original nebula had, the final system has to have exactly the same amount. Numerically, angular momentum is the speed of rotation multiplied by the radius – so when the radius gets smaller, the speed has to increase. This is more intuitive than it sounds – if you start yourself spinning (e.g. on a swivel chair, as in the picture opposite) you speed up when you shrink the radius by drawing your arms in. Because the nebula began life spread out over a huge volume of space, even a very slow initial rotation translates to a much faster rate when the radius shrinks down to the size of a star.

The problem is stars don't spin fast enough to account for all the angular momentum the nebula should have had. In the case of the Solar System, we know exactly where it went. The planets, although they add up to less than 1 per cent of the Sun's mass, account for an astonishing 99 per cent of the Solar System's angular momentum. And because physics works the same way everywhere, it's likely that planetary

**The conservation of angular momentum makes
you spin faster when you pull your arms in.**
(Wikimedia Commons user MikeRun, CC-BY-SA-4.0)

systems have soaked up the 'missing' angular momentum around other stars, too.

Before we go off looking for exoplanets, it's worth mentioning another point that became increasingly obvious as the science of astronomy developed. Not all stars are the same – there's a whole zoo of them, determined by their mass and the stage they're at in their life cycle. In the present context, we're particularly interested in Sun-like stars, which means focusing on a phase of the life cycle called the 'main sequence'. This is where stars use nuclear fusion to convert hydrogen into helium, radiating the energy released as heat and light.

Counterintuitively, the length of time that a star spends on the main sequence is shorter the higher its mass is. So, since we're interested in exoplanets that have existed long enough for life to take hold, we can safely ignore stars that are much more massive than the Sun. In the jargon of the subject (see box), that means concentrating our attention on F-, G-, K- and M-type main sequence stars. The Sun is a G-type, as is another of the stars we've already encountered in this book: Tau Ceti, one of the pair of nearby stars that Frank Drake originally pointed the Green Bank dish at in search of SETI signals. Epsilon Eridani, his other target, is a smaller K-type star.

Stellar Classification

The standard system for classifying stars dates from the late 19th and early 20th centuries, and – unusually for a scientific achievement of that time – we owe it entirely to women. That's women, plural – a whole team of them, led by Annie Jump Cannon at Harvard Observatory, building on earlier work by Williamina Fleming and Antonia Maury. The system they developed is based on spectral lines (see box, page 40), but in the case of main sequence stars it comes down to a question of mass and temperature. Originally a neatly alphabetical system, some of the letters got dropped, while others fall outside the range we're interested in – so we only have to look at four types: F, G, K and M.

The Sun is a G-type main sequence star. This class ranges from 0.8 to 1.1 times the mass of the Sun, making any G-type star really very Sun-like. Going up in mass we come to F-types, between 1.2 and 1.6 solar masses, and as much as six times as bright as the Sun. These stars are hotter: 6,000–7,000°C, compared to 5,000–6,000°C for G-type. The fact that they're brighter and hotter means that the habitable zone, where planets with liquid water might exist, lies further out than in our Solar System.

The opposite direction takes us to K-type stars, covering the range from 0.5 to 0.8 solar masses. These are cooler than the Sun, by a thousand degrees or so, and dimmer, with luminosities a few tenths that of the Sun. The habitable zone is closer to the star than ours, but K-types have the advantage of being longer-lived, so life has a better chance of taking hold and evolving.

Finally there are M-type stars. These are small – less than half the mass of the Sun – and significantly cooler even than K-types. They're so cool, in fact, that most of their energy is radiated in the infrared rather than the visible part of the spectrum, giving them a dull red colour. This, together with their small size, has resulted in a nickname that will be familiar to fans of classic British sitcoms: 'red dwarf'.

Having been indoctrinated since the 16th century with the idea that there's nothing special about the Sun, it's easy to assume that means it's just an average main sequence star. But that's not true. In terms of mass, luminosity and temperature, it's well above average. G-type and F-type stars – together with all the brighter, shorter-lived types – make up the largest, brightest and hottest 12 per cent of all main sequence stars in the galaxy. The dimmer K-types account for another 12 per cent, leaving a massive 76 per cent for M-types – the red dwarfs. That's what a really 'average star' looks like.

Given that they make up such a huge proportion of the galaxy's stars, you may wonder why you don't see a sky full of red dwarfs at night. The reason is they're too dim too see with the naked eye. That's even true of red dwarfs in our own backyard, such as Proxima Centauri – at 4.2 light years the closest other star to the Sun – or the similar-sized Barnard's Star, a little further away at 6 light years.

Could an exoplanet orbiting a red dwarf – a star that's completely different from the Sun – ever harbour life? It's an intriguing question, and one that we'll come back to at the end of this chapter. Before then, there's a more pressing problem to deal with. Even if we look at Proxima Centauri through the world's most powerful telescope – or at a much larger nearby star, such as the F-type Procyon, 11 light years away – all we see is a tiny pinpoint of light. So what hope is there for finding a planet orbiting such a tiny dot?

How to Detect Something You Can't See

As it happens, there's a classic problem in astronomy that's conceptually quite close to detecting an exoplanet – and solutions to that classic problem have been around for a long time. It has to do with binary systems, consisting of two stars orbiting around their common centre of gravity. The situation appears to depend on stellar mass, with F-type and brighter stars often being part of such a system, Sun-like stars occasionally so, and red dwarfs much less frequently. On balance, it's believed that around a quarter of the galaxy's stars belong to binary systems.

In some cases the stars in a binary are far enough apart that they can be seen separately through a telescope, but that's not always the case. So astronomers had to work out how to detect the presence of a binary companion when they couldn't actually see it. They came up with three main methods – all of which are potentially applicable to exoplanet detection too.

The starting point for all three methods is the orbital motion of the stars. Because they're orbiting around a common centre of gravity, both the stars are moving – even if one is much more massive than the other. Suppose a bright star has a smaller companion that's too dim, or too close to the other star, to be seen through a telescope. As it travels around its orbit, it will cause the larger star to 'wobble' slightly from side-to-side – and that wobble may be visible in a sequence of photographs taken at different times. That's the first way that an unseen binary companion can be detected, and it's called the 'astrometric method'.

The second approach to finding a binary star is something you can try for yourself, with no telescope or photographic equipment required. Well, you can in the case of one particular binary star, anyway. It's called Algol, and it's the second brightest star in the constellation Perseus (which can be seen high in the sky on winter evenings in the northern hemisphere). The star's name comes from the Arabic for 'the ghoul', after its distinctly spooky behaviour. If you observe it for several nights running, you'll notice that every three days or so it suddenly becomes dimmer for a few hours. That's because it's a close binary, and you're normally seeing the light from both stars – except for the brief time that one of them passes in front of other, when the light dims dramatically.

We've already encountered the idea of a star's 'light curve' – fluctuations in its brightness over time – in the context of alien megastructures in chapter 4. The term originated in the more sober context of variable stars like Algol – and it gives us another useful tool in the hunt for exoplanets.

The third technique is the most convoluted. Thinking back to the astrometric technique, and the idea of a side-to-side wobble, what about the corresponding motion in the perpendicular direction – towards and away from us? When I was a youngster, long before the days of cheap, radio-controlled drones, the closest thing we had were 'control-line' model aircraft. These were attached to long wires, used by the operator to fly the plane round and round in circles. That person heard a steady buzz from the engine, but anyone standing some distance away heard a completely different sound, constantly rising and falling in pitch.

This, of course, is an example of the Doppler effect, which makes an engine sound higher pitched when it's approaching and lower when it's moving away. That's not just a property of sound, but of all kinds of waves. They get bunched up – hence shorter wavelength and higher frequency – when they're pushed towards you, and stretched out – longer wavelength and lower frequency – when they're pulled away. Exactly the same thing happens to the light from a binary star. It's shifted to longer wavelengths, or redder colours, when the source is receding in its orbit, and to shorter, bluer, wavelengths when it's approaching.

The problem is, you can only make use of this if you know what the wavelength was to start with, so you can work out how much it's been shifted. Here we can draw on an idea we met earlier in the book. The concept of spectral lines cropped up back in Chapter 3, when we were talking about characteristic radio emissions from neutral hydrogen gas. At the much higher temperatures found in stars, similar narrowband lines are produced at visible wavelengths. These include both emission lines – which are brighter than the background spectrum – and absorption lines, which are darker. Measuring the shift in these lines allows us to determine the orbital speed of a binary star – and because it involves looking at the star's spectrum, it's called the spectroscopic method.

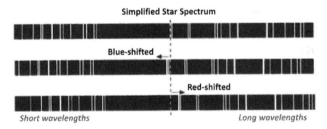

Simplified Star Spectrum

Blue-shifted

Red-shifted

Short wavelengths   Long wavelengths

The characteristic lines in a stellar spectrum are shifted to the blue when the star is approaching us, and to the red when it's moving away.

(NASA image)

So let's see how those three techniques – astrometric, spectroscopic and light-curve – apply to the search for exoplanets. The astrometric technique might seem a non-starter, because the side-to-side wobble is small enough even when a companion star is producing it, let alone a planet. But a couple of claims have been made, the first by Captain William Jacob in 1855. A British army engineer and keen astronomer, he was posted to the Madras Observatory in India. While studying the nearby binary system 70 Ophiuchi, consisting of two K-type stars, he thought he detected an additional 'wobble' in one of them that couldn't be explained by the companion star. In his account of the work, he came to the following conclusion:

> There is, then, some positive evidence in favour of the existence of a planetary body in connection with this system, enough for us to pronounce it highly probable.

Unfortunately, it wasn't as clear-cut as that. While a few other astronomers claimed to have observed the same anomalous wobble, most saw nothing that couldn't be explained by the binary system. If there is an additional wobble caused by a planet it's on the very limits of detectability, and even today no one knows for certain if 70 Ophiuchi has a planetary system or not.

Another claim for the astrometric detection of an exoplanet was made by the Dutch-American astronomer Piet van de Kamp in 1963. The parent star in this case was one we've already met: Barnard's Star, the red dwarf that is one of the Sun's closest neighbours. This one turned out to be a 'bad news/good news' situation. The bad news was that, within a few years, other astronomers had shown van de Kamp's measurements to be the result of instrumental errors. The good news is that in 2018 Barnard's Star was found to have a planet after all. We'll come back to it later in this chapter but before then let's have a closer look at the method that was used to find it. This was the spectroscopic technique, looking for Doppler shifts in a star's spectrum as the gravitational pull of an orbiting planet causes it to wobble.

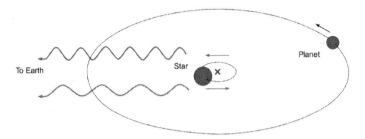

As a planet orbits around a star, the star itself wobbles slightly – and this causes changes in the wavelength of the light it emits due to the Doppler effect.

(Public domain image)

As neat as this idea is, it's not an easy thing to observe. One problem is that a star is so much more massive than a planet that the speed variation we're looking for is very small. The largest planet in our Solar System, Jupiter, causes the Sun to wobble at a mere 12.5 metres per second (around 45 kilometres per hour), which is hardly an 'astronomical' number. Nevertheless, by the 1990s spectroscopic techniques were sensitive enough that this sort of variation was potentially detectable.

The second problem is with the timescale, rather than the size, of the speed variation. A wobble can only be detected by comparing spectra taken at different times, which can mean a very long wait if the rate is low. Jupiter, for example, takes almost 12 years to complete an orbit around the Sun, so astronomers would need to acquire at least that much data before they could be certain of the wobble. The Earth goes round the Sun more quickly – once a year – but it's much smaller than Jupiter, so we're back with the size problem again.

There are two reasons why the Earth takes less time to complete an orbit than Jupiter. Firstly, it has a shorter distance to travel, because it's closer to the Sun so its orbit has a smaller circumference. Secondly, it's travelling at a higher speed. That's because the pull of gravity is stronger at shorter range, so the planet has to whizz round more quickly to keep from falling into the Sun. The result is another 'square–cube law' – though no relation to the one we met in Chapter 2. This one says that the square of the time needed to complete an orbit is proportional to the cube of the average radius – a result known as Kepler's third law, after the 17th-century astronomer Johannes Kepler (whose 21st-century namesake we're going to meet very shortly).

If you've followed me so far, you won't be surprised by what I'm going to say next. When the first exoplanet was discovered by the spectroscopic method, in 1995, it turned out to be very large – comparable in size to Jupiter rather than the Earth – and orbiting very close to its parent star. The astronomers who found it, Didier Queloz and Michel Mayor of the University of Geneva, shouldn't have been surprised either – but they were.

The planet wasn't just close to its parent, a G-type star called 51 Pegasi (51 Peg for short), it was almost unbelievably close. Mercury, the innermost – and smallest – planet in the Solar System, at about 50 million kilometres from the Sun, takes 88 days to complete an orbit. By comparison, 51 Peg's Jupiter-sized planet is a mere 8 million kilometres from it, and completes an orbit every 4.2 days. Astronomers had no idea there could be planets with a 'year' as short as that.

Queloz and Mayor's planet was given the designation 51 Peg b, based on the logic that the star itself was discovered first, so it's tacitly 'a', and any subsequent planets that might be found would be assigned the letters c, d and so on. That's become the standard naming convention for all exoplanets.*

Since the discovery of 51 Peg b, several hundred further exoplanets have been discovered by the spectroscopic method. Many of them, particularly in the earlier days, similarly turned out to be large planets orbiting very close to their parent star – a class that has come to be known as 'hot Jupiters'. That's not to say such planets are especially common, but they're just the easiest to detect by this method. In any case, they're not much use to us from an astrobiological perspective, because they'll be far too hot for any kind of 'life as we know it'. Fortunately, spectroscopic techniques now have a resolution of better than one metre per second, which means they're capable of detecting Earth-like planets as well.

Despite its successes, the spectroscopic method has been overtaken in the planet-hunting stakes by the so-called 'transit method'. This uses a star's light curve in a way that's

---

* For some masochistic reason, the International Astronomical Union has since started giving proper names to exoplanets – for example 51 Peg b is officially known as Dimidium. With hundreds of new exoplanets discovered each year, that's going to get very tedious very quickly.

analogous, but not quite identical, to the detection of binary stars like Algol. In that case, we normally see the light from two stars, with periodic dips in the curve when one star passes in front of the other. But planets don't produce any light of their own, so the best we can hope for is a tiny dip when the planet passes in front of the star – a situation known in astronomical jargon as a 'transit'.

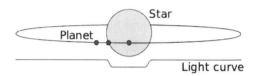

A star's light curve shows a tiny dip when a planet passes in front of it.

(Wikimedia Commons user Nikola Smolenski, CC-BY-SA-3.0)

To put it in perspective, if astronomers in another star system measured the light curve of our own Sun, they would see a dip of about 1 per cent during a transit of Jupiter, and just 0.01 per cent during a transit of Earth. As small as that sounds, even such a minuscule variation can be detected by specially designed instruments – of which the most famous was the Kepler space telescope. Launched by NASA in 2009, it differed in several ways from the archetypal space telescope, the Hubble. While the latter is an ordinary satellite, orbiting the Earth, Kepler was put into orbit around the Sun. Its orbit is very similar to the Earth's own, but at a slightly greater distance – by about 1.3 per cent.

Having mentioned Kepler's third law (that's the other Kepler, after whom this one was named) a few pages ago, you can see what that means. If the orbit is 1.3 per cent larger, it takes 1.2 per cent longer to go all the way round it. If you like playing with squares and cube roots on a calculator, you can confirm that for yourself – otherwise just take my word for it. In any case, the exact number isn't important. The point is that Kepler gradually lags further and further behind the Earth as it travels around the Sun. The advantage of that is that the telescope gets a completely uninterrupted view of the sky, without the Earth repeatedly getting in the way – as it would if it was orbiting around it.

The fact that Kepler has its own private orbit around the Sun has another consequence. Imagine there was a giant telescope inside the Earth, pointing out of a hole at the North Pole. As you're probably aware, the North Pole always points in the same direction in space, towards a star called Polaris – alias the Pole Star – in the constellation of Ursa Minor. It was the same with Kepler. The telescope constantly pointed at the same part of the sky, in this case in the direction of the constellations Cygnus and Lyra. That meant it could look very deeply at all the stars in this direction – far more than could be seen with ordinary telescopes – out to a distance of 3,000 light years or more.

There's one final difference between Kepler and a more conventional telescope like Hubble. The latter is equipped with numerous instruments for different tasks – such as the spectacular high-resolution photography for which it's so famous. Kepler, on the other hand, just had one single purpose: to record stellar light curves, with the aim of detecting exoplanets. By the time it was retired in 2018, Kepler had measured the light curves of no fewer than half a million stars.

That was just the start of the process, though. The real hard work involved sifting through the data looking for the characteristic signatures of exoplanets. Much of this was done by computers, but some of it was farmed out to Tabetha Boyajian's Planet Hunters – the citizen science team that discovered a possible alien megastructure (or more likely, huge dust cloud) around Tabby's Star, as described in Chapter 4. Unlike that other citizen science project, SETI@home, which just borrowed time on home computers, Planet Hunters genuinely needed human input – because people are better than computers at spotting subtle patterns in raw data.

By the time Kepler retired, all this effort had yielded at least 2,800 new exoplanets – with at least as many 'possibles' still waiting to be confirmed. If you want to, you can take a look at the data yourself and see if you can find a planet everyone else has missed – because NASA made the whole dataset freely available to the public in November 2018.*

Kepler was one of the comparatively few NASA missions in recent years that succeeded in capturing the imagination of the public. The main reason, of course, was its nine-year mission to seek out strange new worlds – but I suspect that having a catchy name helped too. I mean, 'Kepler' is so much better than, say, 'FRESIP', isn't it? Yet FRESIP – short for Frequency of Earth-Sized Inner Planets – was exactly what the mission was called in the original proposal, before the PR people got onto the case.

On the other hand, when Kepler's successor was launched in April 2018, it made it all the way into space with

---

* https://archive.stsci.edu/kepler/

its original acronym intact: TESS, for Transiting Exoplanet Survey Satellite. That's slightly more imaginative than FRESIP, but it remains to be seen if TESS will generate as much enthusiasm as Kepler did.

Actually, the TESS mission is potentially even more exciting than Kepler. Although it uses the same transit method to detect exoplanets, there are some important differences. Kepler looked at hundreds of thousands of stars, all in the same small part of the sky – which meant that most of them were a long way away. On the other hand, TESS – in a more conventional orbit around the Earth – will look at the whole sky, but concentrate mainly on nearby stars. It's expected to find at least as many exoplanets as Kepler did, but far more of them are going to be in our backyard in galactic terms. Discovering a habitable world that may contain life will be that bit more exciting if it's only 20 light years away, rather than 20,000.

Signs of Life

As we saw in the previous chapter, two of the key requirements for Earth-type life – a source of energy and a stockpile of essential elements – are not hard to come by. The third prerequisite, liquid water, is a different matter, because it only exists over a comparatively narrow range of temperatures. This leads to the idea of a 'habitable zone' around a star – the range of orbits where there's enough heat to melt ice, but not enough to vaporise it completely. The Earth, of course, at 150 million kilometres from the Sun, lies in the

Solar System's habitable zone. On the other hand Venus, 30 per cent closer, and Mars, 50 per cent further away, are outside it.

The situation will be similar in planetary systems around other G-type stars, while the scale shrinks in the case of cooler K- and M-types, and expands for F-type stars. In the latter case there's another factor to take into consideration, too. Earlier, it was mentioned that the lower temperature of red dwarfs means that they radiate a lot of their energy below the visible part of the spectrum, in the infrared. The opposite is true with an F-type star. Because it's hotter, a greater fraction of its energy is radiated *above* the visible spectrum, in the DNA-damaging ultraviolet. With the usual caveat that we're talking about 'life as we know it', that suggests that F-type stars are less likely to have habitable planets than cooler stars.*

Even our own Sun produces some UV, of course, but we're protected from the worst of it by the Earth's atmosphere. Another thing we're cosily protected from is so-called 'space weather'– a completely different physical effect from UV, but with similarly destructive effects. In essence, space weather is a wind of fast-moving charged particles emitted by the Sun. We're protected from it, not by the Earth's atmosphere, but by its magnetic field, which deflects most of the particles before they reach us. Winds of this type would be very bad news for life on a planet that lacked a magnetic field – especially if it was close to its parent star, as it would need to be in the case of a K- or M-type star.

---

* The fact that F-type stars have shorter lifetimes doesn't help either.

Up to this point, the discussion has been frustratingly hypothetical. *If* an exoplanet is in the habitable zone, and *if* it's protected from harmful radiation and space weather, then it *might* harbour life. But 'if' and 'might' aren't good enough. We want to know for certain whether or not a particular exoplanet has organisms living on it.

This brings us back to the subject of 'biosignatures': indicators of the presence of life that we can detect from a distance of many light years. The most promising approach is through the analysis of an exoplanet's atmosphere, in which context biosignature chemicals are often referred to as 'biomarkers'. We've already seen that the oxygen that makes up 21 per cent of the Earth's atmosphere is basically a waste product of living things, such as bacteria and plants. Another waste gas that's produced by certain bacteria – including the kind that live inside a cow's stomach, helping it to digest grass – is methane. The average dairy cow can burp as much as 500 litres of methane per day into the atmosphere.

Methane is only a very minor constituent of the atmosphere – a tiny fraction of 1 per cent – but the point is it wouldn't be there at all if it wasn't constantly replenished by biological processes. Left to itself, the methane would quickly combine with oxygen to form carbon dioxide and water. Oxygen, in fact, likes to combine with a whole range of things, so it too would disappear in short order if it wasn't being renewed somehow. In principle there are other, non-biological processes that could create fresh oxygen and methane, but they're not very likely, so these two gases remain the best biomarkers we know of.

All we need now is a method of measuring the composition of an exoplanet's atmosphere. Fortunately there's an ingenious way to do just this, first applied in 2000 by Sara Seager – then of the Institute for Advanced Study in Princeton, now of the Massachusetts Institute of Technology – to a planet with the unmemorable name of HD 209458 b. This is another 'hot Jupiter', orbiting a G-type star at about one-twentieth of the Earth's distance from the Sun. If you think that sounds reminiscent of 51 Peg b, you won't be surprised to learn that HD 209458 b was discovered by the same spectroscopic method – but it was subsequently observed using the transit method too. In fact, because its orbit is so small, transits occur very frequently – which is a good thing, because they're the key to Seager's method.

When the planet is behind the star – technically called a 'secondary transit' – it's completely hidden from our view. So all we see in the star's spectrum are the emission and absorption lines characteristic of the star's own chemical make-up. On the other hand, when the planet is passing in front of the star – a primary transit – we see additional absorption lines due to the light from the star passing through the planet's atmosphere. All Seager had to do was subtract the secondary transit spectrum from the primary transit one to get a spectral signature for the planet alone.

In itself this was a fascinating piece of science, but it's not much use in astrobiological terms because a hot Jupiter like HD 209458 b – with a surface temperature of more than 1,000°C – is the last place we'd expect to find 'life as we know it'. The really interesting thing will be to apply the same technique to smaller, Earth-like exoplanets lying in the habitable zone of their parent stars. The problem is, there aren't any instruments sensitive enough to do that at the moment.

Fortunately there's one in the pipeline. Called the James Webb Space Telescope, or JWST, it's NASA's successor to the Hubble. Already in an advanced stage of construction, and scheduled for launch in 2021, the JWST should be a huge improvement on its venerable predecessor. It will be much larger, for one thing, with a main mirror 6.5 metres in diameter compared to the Hubble's 2.4 metres. The new telescope will be much further from Earth, too – about 1.5 million kilometres away – which should give it a clearer view of the sky. It won't strictly be in Earth orbit, or in an orbit around the Sun like Kepler, but a sort of cross between the two – at one of the so-called Lagrangian points, where the gravitational pull of the Sun and Earth effectively cancel each other out.

It's fair to say that the JWST hasn't had a smooth ride, even before it's been launched. Ever since it was conceived, in the first years of this century, the project has been subject to repeated delays and cost overruns. The latter are particularly controversial, because they drain money from NASA's overall astronomy budget. In 2010, the magazine *Nature* ran an article on the JWST called 'The Telescope that Ate Astronomy'. The piece refers to a scheduled launch in 2014 – a date that was optimistic by at least seven years.

When the JWST finally gets into space, it will be used for a whole range of projects – but one of the most exciting is being planned by Sara Seager and her research group. They are going to use the JWST to analyse the atmospheres of the most promisingly habitable-looking exoplanets detected by the TESS mission – which should be coming to an end around the time JWST goes into space.

What are Seager's chances of success? That's rather like the question everyone was asking Frank Drake when he started listening for SETI signals – and came up with the Drake equation by way of an answer. Seager approached the problem in the same way, creating an analogous formula now known as the 'Seager equation'. It uses the same kind of logic as the Drake equation to address a different question: how many planets with biomarker chemicals in their atmospheres could be detected by the JWST/TESS combination, given the observational parameters of those telescopes? Her answer, after crunching all the necessary numbers, is 'at least one'. Of course, that's what Frank Drake said – but Seager's problem is simpler than his, so let's hope that this time it's the right answer.

Other Earths?

At the time of writing (mid-2019), more than 4,000 exoplanets have been found – around two-thirds of them by Kepler alone. That's a big list, and it needs to be narrowed down if we want to zero in on Earth-like planets in the habitable zone. The University of Puerto Rico has done just that, in its online 'Habitable Exoplanets Catalogue', which already contains around 50 promising contenders.

One of these belongs to a star we've encountered already: Tau Ceti, the nearby G-type star that was one of the targets for Frank Drake's original SETI search. It's now known to have at least five planets – one of which, Tau Ceti e, is in the habitable zone. This planet is similar in size to the Earth, and orbits at a smaller distance – compensating for the fact that Tau Ceti is only about half as bright as the Sun. At less than 12 light years, it might just be Earth's closest inhabited neighbour.*

Then again, there are a couple of other contenders for that title. They're more controversial because they belong not to G-type stars, or even K-type, but the much dimmer M-type – alias red dwarfs. We've already met the two we're talking about: Proxima Centauri, the nearest other star to the Sun, and Barnard's Star, the fourth nearest. Despite their closeness – around 4 and 6 light years respectively – the two stars are so faint they're not even visible to the naked eye. That doesn't prevent them having a habitable zone, but it means it's located much closer to the star than the Earth is to the Sun.

The possibility of habitable planets existing around red dwarfs is intriguing for a couple of reasons. The first is that these stars are so common, making up more than three-quarters of all the main sequence stars in the galaxy. Secondly – and more importantly – the fact that M-type stars have incredibly long lifetimes means that, on average, they're going to be significantly older than the Sun. If they've been around longer, that means life on any suitable planets orbiting them will have had more time to evolve into advanced – and possibly highly intelligent – forms.

---

* The other nearby star targeted by Drake, Epsilon Eridani, also has an Earth-sized planet. However, it's much too far out – from a star that's only a K-type to start with – to be habitable.

There are problems, though. Because M-type stars are cooler than other types, a larger proportion of their energy is radiated at infrared rather than visible wavelengths. That makes photosynthesis, the process by which terrestrial plants extract energy from sunlight, much more difficult – though not totally impossible.

Other problems arise simply because the planet would have to be located so close to the star. The effects of space weather, for example, would be much more severe for a close-in orbit. Unless the planet is protected by a very strong magnetic field, estimates suggest the constant stream of high-energy particles could strip its atmosphere off completely. That, needless to say, would be bad news for any lifeforms trying to evolve there.

Another effect that becomes much stronger at short ranges is the tidal force: the differential pull of gravity on opposite sides of a planet. Here on Earth, for example, the tidal effect of the Moon is greater than that of the Sun, despite the fact that the latter is so much bigger. That's because the Moon is a lot closer to us. The Moon also feels a tidal pull from the Earth – and because the Earth is bigger, it's even stronger. Over time, the Earth's tidal pull has slowed the Moon's rotation to the point that it always keeps the same face pointing towards us. This phenomenon – called 'tidal locking' – is also expected to occur with some of the close-in planets orbiting around red dwarfs. It would mean the side pointing away from the star was permanently cold and dark – but it wouldn't necessarily affect the habitability of the side facing the star.

Tidal forces have another effect – one that we encountered in the previous chapter. Remember how the tidal pull of Jupiter is thought to produce an internal heating effect inside moons like Europa? It could be enough to turn a world that ought to be frozen solid into a potential habitat for life. In the case of a planet orbiting close to a red dwarf, the same effect would provide an additional heat source on top of the heat radiated from the star itself. Taking that into consideration would shift the habitable zone out to a somewhat greater distance from the star.

Now let's return more specifically to our two red-dwarf neighbours. To start with, Proxima Centauri has one known planet, discovered by the spectroscopic method in 2016. At first sight it looks very promising. It's listed in the Habitable Exoplanets Catalogue because it's a similar size to the Earth and at a suitable distance from the star for liquid water to exist on its surface. There's a serious problem, though. Unlike the Sun, which radiates light and heat at a steady rate, Proxima Centauri is classed as a variable star. Every few months, its output flares up – sometimes to many times its normal level – before dying back down again a few minutes later. If our own Sun did a similar thing, even once, the sudden blast of radiation would probably kill off all but the lowest forms of life on Earth.

That doesn't sound very promising, so let's shift our attention to the other nearby red dwarf, Barnard's Star. That's the one where Piet van de Kamp claimed to have discovered an exoplanet in 1963, before other astronomers decided it was just an instrumental error – and then in 2018 a planet was discovered there after all. It was given the inevitable, if rather clumsy, name of Barnard's Star b. It's similar in size to the Earth, but orbits quite a way outside the star's habitable zone. At first sight that ought to mean it's frozen solid – but the possibility that it's subject to Europa-style tidal heating means it might actually be hospitable to life after all.

In addition to those two very close ones, several other promising exoplanets have been found orbiting red dwarfs. I won't bore you by cataloguing them all here, but three of them are definitely worth mentioning – because they all happen to belong to the same star.

The star in question is the famous TRAPPIST-1 – so named because it was discovered, in 2016, by the Transiting Planets and Planetesimals Small Telescope, or TRAPPIST for short. The telescope is located in Chile, but it's operated by a team of Belgian astronomers. Now Belgium is famous for a lot of things, but the most important* is the very strong beer brewed in the Trappist monasteries there. That's an extraordinary coincidence, because 'Transiting Planets and Planetesimals Small Telescope' is such an obviously uncontrived name it can't possibly have been influenced by its operators' drinking preferences.

Even by red dwarf standards, TRAPPIST-1 is tiny, with a diameter only about 20 per cent larger than the planet Jupiter. But it's much denser than Jupiter, and it shines with its own light – albeit 2,000 times less brightly than the Sun. Despite its small size, TRAPPIST-1 has a retinue of at least seven planets, with no fewer than three of them – e, f and g – lying in the habitable zone. Whether that means they're actually inhabited is another matter, though, since the star is so completely different from anything we're familiar with. The

---

* Yes, this is a subjective judgement.

fact is we need to find out much more about TRAPPIST-1's planets than the handful of facts and figures – diameters, orbital distances and masses – that we already have.

This dearth of information may be frustrating to scientists, but it proved to be no impediment to artists – who promptly came up with some stunning visualisations of what the TRAPPIST-1 system 'might' look like. The one shown here was created for NASA by Robert Hurt and Tim Pyle, and depicts a conjectural view from TRAPPIST-1f – with two of its neighbouring planets faintly visible in the sky as the dim red sun sinks below the horizon.

Whether there are any lifeforms on TRAPPIST-1f to witness that dramatic sunset is something we simply don't know yet. The fact is, despite the thousands of exoplanets we've discovered over the last couple of decades, we've yet

An imaginative visualisation of the landscape on TRAPPIST-1f, based on the tiny amount of factual information available.

(NASA image)

to find conclusive evidence of life on any of them. What happens when we do – and I'm sure it really is 'when', not 'if' – is a subject we'll address in the next chapter.

FIRST CONTACT 7

Although speculations on the existence of extraterrestrial life go back centuries – including many based on scientifically sound arguments – the serious science of astrobiology only started around 60 years ago. By 'serious science' I mean specific theoretical predictions of the indicators to look for – such as the infrared signatures of Dyson spheres, or artificial-looking signals at the radio astronomer's favourite frequency of 1420 MHz – and the methodical searches that followed. Then again, can it really be called serious science if, even after 60 years, there's still no hard evidence that the subject it purports to study actually exists?

As it happens, there are precedents for this situation. The Higgs boson, for example, was originally postulated in 1964 and soon became part of the standard model of particle physics, even though it wasn't actually discovered until 2012. There was an even longer delay – a whole century – between Einstein's prediction of gravitational waves in 1916 and their eventual detection in 2016. But they too were part of the accepted worldview of the physics community, which is why most scientists were pretty confident the discovery would be made in the end.

The existence of extraterrestrial life is almost – but not quite – in the same category. It's not a completely unavoidable consequence of the laws of nature, but it's pretty close. We believe life arose on Earth as a result of processes that ought to be duplicated on other similar planets, and there's absolutely no evidence to indicate that the Earth, or the Solar System, is unique or special in any way. Based on that line of argument, the discovery of extraterrestrial life is – just like the Higgs boson and gravitational waves – simply a matter of time.

Over the years there have been tantalising hints. As we have seen, when NASA's Viking landers scooped up some Martian soil in 1976, one of the three tests designed to find evidence of life produced a positive result – but the other two didn't. With the benefit of hindsight, scientists have since thought up plausible non-biological explanations for that one positive result. Then a year after Viking, the Wow signal had many of the characteristics expected of an interstellar radio beacon – but it was tantalisingly brief, and has never been repeated.

Fast forward to August 1996, and NASA scientists came as close as they ever have to announcing the discovery of alien life – supposedly found in fossil form in the Martian meteorite ALH84001. But the announcement didn't go down well with other scientists, who quickly pointed out numerous alternative explanations for the various pieces of evidence the NASA team had put forward.

More recently still, the discovery of the anomalous light curve of Tabby's Star in 2015 was tantalisingly suggestive of the presence of an alien megastructure – although it seems more likely now that it was merely a cloud of dust. And the mysterious object dubbed 'Oumuamua, which passed through the Solar System on an interstellar trajectory in 2017, might – just might – have been an ancient alien space probe.

In all these cases, while they created a flurry of excitement at the time, the general scientific consensus has come round to the view that there's a more conventional explanation which doesn't require the hypothesis of extraterrestrial life. That's not to say, however, that the consensus is necessarily right, and the 'life' hypothesis remains an outside possibility. In one case – that of 'Oumuamua – we'll never know the real answer, because we're never going to see it again. But in all the others, there's the possibility of obtaining further data which could shift the odds again.

As far as life on Mars is concerned – either now or in the distant past – the first serious attempts to move beyond the Viking era should be made in 2021, with the arrival of two rovers, from NASA and ESA, both with a specific astrobiological remit. Not long after that, NASA's Europa Clipper and ESA's JUICE spacecraft should blast off in search of life on Jupiter's moons. As regards more distant worlds that might harbour life, the TESS space telescope is continuing its search for exoplanets, and the most promising of these will be passed to the James Webb Space Telescope to look for atmospheric biosignatures.

All those missions are looking for relatively low-end indicators of alien life. At the other end of the scale, the kind of

SETI searches that turned up the Wow signal are still going on – these days with far more powerful radio telescopes. Of particular note is the Breakthrough Listen initiative – which, with its multi-million-dollar budget provided by Yuri Milner, is able to devote far more time to SETI activities than previous surveys. And now that the discovery of Tabby's Star has raised awareness of another kind of technosignature, astronomers have started searching through stellar light curves for other anomalies that might indicate the presence of an alien megastructure.

Unless there's something very wrong with our picture of the universe, it's only a matter of time before we finally find evidence of extraterrestrial life. Thanks to science fiction, we all have a clear mental image of what 'first contact' will mean. There will be a high-profile press conference, indisputable evidence will be presented, the whole world will cheer and dance in the streets, and life on Earth will never be the same again. Unfortunately, the reality is likely to be much more low-key than that. To see why, let's run through a few scenarios.

Possible Futures

If we think about the astrobiological discoveries that might be made over the next ten to twenty years, one of the most likely is the discovery of biomarkers in the atmosphere of an exoplanet. In fact, based on current expectations, this is highly probable – as Sara Seager showed with her biosignature version of the Drake equation. But it won't lead to a dramatic, overnight announcement. Scientists are hyper-cautious, and they're going to put weeks, or months, of effort into analysing and interpreting the data, and looking for any other processes that might give a 'false positive' result. Even after the team announces the 'discovery' of life on an exoplanet – based, say, on the presence of oxygen and methane in the atmosphere – there will be plenty of other scientists ready to play devil's advocate by thinking up non-biological interpretations of the data.

It also seems highly possible that evidence of some form of alien life – almost certainly microscopic – will be found much closer to Earth, say on Mars or Europa. But it's still likely to take the form of a gradual realisation and acceptance, rather than the sudden world-shaking announcement movie-goers have been led to expect. In the end, however, it may be less ambiguous, because it will be easier to make follow-up observations with other instruments. Multiple independent indicators of biological activity will be more convincing than just a single piece of evidence.

As fascinating as these possibilities are, they won't change the world – because ultimately they're simply confirmatory of what scientists already believe. Of course, they'll make huge headlines, in a way that scientific discoveries rarely do, and those headlines will almost certainly talk about a 'revolution'. But it will be a misuse of the word, because a real revolution would involve changing the way we look at the world. Discovering microbes on Europa, or biomarkers in the atmosphere of an exoplanet, won't do that. Like the Higgs boson and gravitational waves, it will simply confirm that we've been thinking along the right lines. Within a few months, most people – scientists and non-scientists alike – will have forgotten there was ever any doubt about the existence of extraterrestrial life.

For astrobiologists, that first discovery – whatever form it takes – will be a huge boost. It will remove any last lingering suspicion that their subject isn't a 'real' science – and will undoubtedly spur far larger and better-funded searches in the future. As to the public, however, I can't help suspecting they'll feel a little cheated. Despite the fact that journalists – even the most sensational kind – usually make it clear in the small print that astrobiology is chiefly concerned with very primitive forms of life, I'm not convinced the message has really sunk in. Certainly the kind of people who get their science education from blockbuster movie franchises – or UFO websites – are unlikely to be too excited by the discovery of the alien equivalent of bacteria.

What a lot of people – and I admit I'm one of them – are really hoping for is evidence not of alien life per se, but *intelligent* alien life. That's what the sci-fi cliché of the world-changing press conference and dancing in the street is all about. It's a discovery that's less likely than microscopic life – for the obvious reason that, almost by definition, there must be far more primitive life in the galaxy than highly evolved intelligence. But the relative likelihood isn't as straightforward as that, because signs of intelligent aliens – such as radio signals, starship drives or megastructures – might be detectable at much greater distances than more primitive biosignatures.

It's a question that was examined in a 2018 paper by Manasvi Lingam and Avi Loeb – the same Harvard astrophysicists who proposed the 'starship drive' theory to explain fast radio bursts. As we established in Chapter 4, they're not just cranks who cry 'aliens' every time a mystery crops up. In this paper, for example, they use rigorous scientific logic – based on generalised versions of the Drake and Seager equations – to estimate the relative likelihood of detecting biosignatures or technosignatures. Needless to say, the answer they arrive at isn't a precise one – but in very broad terms they conclude that detecting technosignatures is around a hundred times less likely than detecting biosignatures. That means the SETI devotees among us have a harder task than the microbe-hunters – but not an impossible one.

Sadly, even then a positive result may not turn out to be the world-changing event it's imagined to be. Think about it. Suppose astronomers discover a star with a light curve that is exactly what would be expected from a Dyson sphere or some other megastructure, and isn't easily explained by any other hypothesis. Yes, it's highly suggestive of aliens – but it's not definitive proof. The same would be true of, say, a mysterious blast of energy that doesn't match any known astrophysical process, but might just be the signature of an interstellar space drive. It would be exciting, but frustratingly inconclusive* – and there would doubtless be debunkers ready to come up with alternative natural explanations.

A structured message, on the other hand, would be much less ambiguous. Even if we can't decipher what it says, it's not that difficult to distinguish an artificial signal from a

---

* Unless, of course, we saw both these things – the megastructure-like light curve and the mysterious blast of energy – coming from the same star system. That would be far more persuasive.

natural one. So that brings us all the way back to where the science of astrobiology started, with Frank Drake's original SETI search and its modern-day counterparts – which are now looking for modulated laser signals as well as radio messages. It's still the most likely way we're going to detect alien intelligence, if the issue isn't going to be surrounded by a huge cloud of uncertainty. It's sufficiently likely, anyway, that people have given serious thought to what will happen if and when we do.

In 1989, the International Academy of Astronautics codified such thoughts in its 'Declaration of Principles for Activities Following the Detection of Extraterrestrial Intelligence'. This isn't a document with any legal power, but it's been ratified by a number of organisations including the International Astronomical Union and the SETI Institute. On the latter's website, Seth Shostak summarises the declaration as follows:

> There are really only three important components to this two-page text. First, the detection of alien life should be carefully verified by repeated observations. Second, the discovery should be publicised. Third, no response should be sent without international consultation.

The implication is that those things are supposed to happen in that order: (1) verify, (2) publicise, (3) consult over possible response. As simple as that sounds, it's impossible in practice. The team that makes the initial discovery can't verify it on their own – they need to call on third-party assistance (in science, that's what 'verification' means) – and they can't get that assistance without making the discovery public. The actual process would be no different from the standard one followed by astronomers after any new discovery – which involves pulling in as many other astronomical resources as possible, as quickly as possible. As the SETI Institute's founder Jill Tarter put it:

> We'll want everybody who can to look at it right away. We'd like people to look in the signal's direction, with different tools, checking different frequencies, and try to figure it out.

Once the news is in the public domain, it's equally impossible to enforce the third part of the declaration, about not sending a response without international consultation. It's an eminently sensible suggestion, because no matter how innocuous-looking the message is, we have no idea what intentions lie behind it. I mean, you wouldn't reply to an email telling you that you'd won a lottery you didn't enter – and by the same logic perhaps we shouldn't answer an interstellar message offering to tell us the secrets of the universe. But there are something like 8 billion people on the planet, and once the frequency and source of the signal have been made public, you can be sure that at least one of those people will try to answer it. All it requires is a large radio dish and a bit of technical know-how.

Of all the scenarios with any likelihood of happening, the receipt of an alien message comes closest to that sci-fi vision of a 'world-changing event'. Whether or not we choose to reply – and whether or not we ever succeed in deciphering it – the very existence of a message would be conclusive evidence that we're not alone in the galaxy. It wouldn't have the ambiguity and uncertainty that comes with other putative technosignatures, and we could be sure that it originated with an intelligent, technologically savvy species, and not – as with more basic biosignatures – just some primitive microorganisms.

Viewed from a purely scientific angle, we could still say it's just 'confirming what everyone expects', like gravitational waves or the Higgs boson. But taken in a broader perspective, it's much bigger news than that. Living in the more secular West, it's easy to forget that a large fraction of the world's population has a non-scientific worldview, which is shaped instead by religious or other non-materialist philosophies. And they're going to be just as excited by the discovery of intelligent aliens as anyone else.

There's a misconception in some circles that the existence of extraterrestrial intelligence would somehow 'disprove' religion. A particularly extreme example can be found in Arthur C. Clarke's novel *The Fountains of Paradise* (1979). In one of its subplots, intelligent radio signals are received from an alien robotic probe – dubbed Starglider – as it passes through the Solar System. One of the novel's characters remarks that 'Starglider has effectively destroyed all traditional religions' – and Clarke himself, in his authorial voice, adds that 'it had put an end to the billions of words of pious gibberish with which apparently intelligent men had addled their minds for centuries'.

Clarke seems to have been under the impression that religious faith depends on the cosy medieval notion that the universe was specifically created for humans on Earth. As we saw in Chapter 2, this idea was old hat by the 17th century, and aliens were enthusiastically absorbed into the religious worldview. Nevertheless, not everyone seems to be aware of this. A survey in 2011 found that a majority of atheists still believe the discovery of aliens would spell the end of religion – although religious believers themselves said it would make no difference at all. In 2015, the then-director of the Vatican observatory, Father José Funes, was widely reported as saying: 'if there was intelligent life on another planet, I don't see that as a contradiction with the Christian faith' – and most other religions around the world take a similar view.

We can't really end this book without touching on the ultimate sci-fi scenario, of face-to-face contact between humans and extraterrestrials. Perhaps the archetypal depiction of this comes in the final scene of the movie *Star Trek: First Contact* (1996). An expectant crowd gazes in awe as an advanced – but not unimaginably advanced – spaceship descends to Earth, the hatch slowly opens, and an alien – almost exactly humanlike, except for the pointed ears – steps out, raises a hand in greeting and says, in flawless English, 'Live long and prosper'.

To cut a long story short, it won't happen like that. The alien won't look remotely human, because it will be the product of a completely different evolutionary tree on a completely different planet. Whatever method it uses to get here, it won't involve anything we would recognise as a spaceship. That's because, given the vast distances involved, the mere fact that it got here means it must be using laws of physics we haven't discovered yet. To be honest, though, the chance that an alien civilisation – at this precise moment in the galaxy's vast history – has both the technology and the motivation to come and visit us is so slim as to be hardly worth considering.

But let's consider it anyway – that point about motivation in particular – as a purely hypothetical question. The one thing we can say with a fair degree of certainty is that if aliens do choose to come all this way to visit us, they'll have a good reason for doing so. Will it be because, as Stephen Hawking feared, they're interstellar Columbuses here to pillage the Earth's resources before moving on to the next planet? Or are they like the Vulcans in *Star Trek*, here to welcome us into a galactic federation, with all the material, technological and intellectual benefits that brings?

Of course, we can have no idea what might motivate alien visitors to Earth. I'd like to think they'd have our best interests at heart, but I'm not convinced. Human actions are rarely driven by interspecies altruism, and as soon as you look at other forms of life on Earth, that 'rarely' becomes 'never'. If the same rule applies throughout the galaxy, we can be grateful for the fact that it was a hypothetical question – and that when 'first contact' does happen, it will be at the safe distance of several light years.

FURTHER READING

Chapter 1: Life Beyond Earth

Ben Miller, *The Aliens Are Coming* (Sphere, 2017)

Andrew May, *Pseudoscience and Science Fiction* (Springer, 2017)

Sean Martin, 'Aliens on Europa: NASA Hunts for Life Just 1 cm under Surface of Jupiter's Moon' (https://www.express.co.uk/news/science/993543/aliens-discovery-europa-jupiter-moon-nasa-news-extraterrestrial-space)

Chapter 2: Thinking About Aliens

Jim Al-Khalili (ed.), *Aliens* (Profile Books, 2016)

Jack Cohen & Ian Stewart, *What Does a Martian Look Like?* (Ebury Press, 2004)

Michael J. Crowe, 'A History of the Extraterrestrial Life Debate', *Zygon*, June 1996

Seth Shostak & Susan Schneider, 'Goodbye, Little Green Men' (https://www.seti.org/seti-institute/news/goodbye-little-green-men-0)

Chapter 3: Extraterrestrial Communication

Paul Murdin, *Are We Being Watched?* (Thames & Hudson, 2013)

Seth Shostak, 'Was it ET on the line?' (https://www.seti.org/was-it-et-line-or-just-comet)

Rebecca McDonald, 'The First Comprehensive, Interactive Tool to Track SETI Searches' (https://www.seti.org/press-release/first-comprehensive-interactive-tool-track-seti-searches)

Laurance Doyle, 'Animal Communications, Information Theory, and the Search for Extraterrestrial Intelligence' (https://www.seti.org/animal-communications-information-theory-and-search-extraterrestrial-intelligence-seti)

Carl Sagan, *The Cosmic Connection* (Hodder & Stoughton, 1973)

Seth Shostak, 'Should We Keep a Low Profile in Space?' (https://www.nytimes.com/2015/03/28/opinion/sunday/messaging-the-stars.html)

SETI@home website: https://setiathome.berkeley.edu/

Chapter 4: Interstellar Engineering

Milan Ćirković, *The Great Silence* (Oxford University Press, 2018)

David Aguilar & Christine Pulliam, 'A New Approach to SETI: Targeting Alien Polluters' (https://www.cfa.harvard.edu/news/2014-21)

Luc Arnold, 'Transit Light-Curve Signatures of Artificial Objects' (https://arxiv.org/abs/astro-ph/0503580)

Nadia Drake, 'Mystery of Alien Megastructure Star Has Been Cracked' (https://news.nationalgeographic.com/2018/01/mystery-of--alien-megastructure--star-has-been-cracked/)

Alex Kasprak, 'What's the Deal With Those Mysterious Fast Radio Bursts from Deep Space?' (https://www.snopes.com/news/2019/01/11/radio-bursts-from-deep-space/)

Robert A. Freitas, 'A Self-Reproducing Interstellar Probe' (http://www.rfreitas.com/Astro/ReproJBISJuly1980.htm)

Marina Koren, 'When a Harvard Professor Talks About Aliens' (https://www.theatlantic.com/science/archive/2019/01/oumuamua-interstellar-harvard-astrophysicist/580948/)

Chapter 5: Starting Small

Athena Coustenis & Thérèse Encrenaz, *Life Beyond Earth* (Cambridge University Press, 2013)

Marc Kaufman, 'Life Found Deep Under Antarctic Ice for First Time?' (https://www.nationalgeographic.co.uk/environment

-and-conservation/2017/11/life-found-deep-under-antarctic
-ice-first-time)

Jonathan Amos, 'Amount of Deep Life on Earth Quantified'
(https://www.bbc.co.uk/news/science-environment-46502570)

Christopher Packham, 'Researchers Produce All RNA Nucleobases in
Simulated Primordial Earth Conditions' (https://phys.org/news/
2017-04-rna-nucleobases-simulated-primordial-earth.html)

Brian Clegg, 'The Physics Theory of the Origin of Life', *BBC Focus*,
February 2019

Andrew Griffin, 'NASA Finds New Form of DNA in Search for Alien
Life' (https://www.independent.co.uk/life-style/gadgets-and
-tech/news/nasa-dna-discovery-alien-life-proof-hachimoji
-detection-a8792636.html)

Mike Wall, 'Life on Venus? Why It's Not an Absurd Thought'
(https://www.space.com/40304-venus-clouds-alien-life-search.
html)

Mindy Weisberger, 'Could Life on Mars Be Lurking Deep
Underground?' (https://www.livescience.com/64318-mars-life
-deep-biosphere.html)

Paul Rincon, 'Europa: Our Best Shot at Finding Alien Life?'
(https://www.bbc.co.uk/news/science-environment-38925601)

Ian Sample, 'Moon of Jupiter Prime Candidate for Alien Life after
Water Blast Found' (https://www.theguardian.com/science/
2018/may/14/europa-moon-of-jupiter-prime-candidate-for-alien
-life-after-water-blast-found)

Chapter 6: Exoplanets

Lucas Ellerbroek, *Planet Hunters* (Reaktion Books, 2017)

Karel Schrijver, *One of Ten Billion Earths* (Oxford University Press,
2018)

Diane Samson, 'All Data Collected By Kepler Now Publicly Available'
(https://www.techtimes.com/articles/235359/20181103/all
-data-collected-by-kepler-now-publicly-available.htm)

Stephen Clark, 'Astronomers Announce First Exoplanets Discovered
by NASA's TESS Mission' (https://spaceflightnow.com/2019/
01/09/astronomers-announce-first-exoplanets-discovered-by
-nasas-tess-mission/)

The Habitable Exoplanets Catalogue (http://phl.upr.edu/projects/ habitable-exoplanets-catalog)

Smithsonian Insider, 'Harsh Space Weather Dooms Life on Red-Dwarf Planets' (https://insider.si.edu/2014/06/harsh-space-weather -may-doom-potential-life-red-dwarf-planets/)

Deborah Byrd, 'Primitive life at Barnard's Star?' (https://earthsky.org/ space/barnard-star-planet-primitive-life-heat)

Chapter 7: First Contact

Manasvi Lingam & Abraham Loeb, 'Relative Likelihood of Success in the Searches for Primitive versus Intelligent Extraterrestrial Life' (https://arxiv.org/abs/1807.08879)

Seth Shostak, 'What Happens Next If We Find Proof of Space Aliens?' (https://www.seti.org/what-happens-next-if-we-find -proof-space-aliens)

Tim Folger, 'Contact: The Day After', *Scientific American*, January 2011, pp. 40–5

Martin Dominik & John Zarnecki, 'The Detection of Extraterrestrial Life and the Consequences for Science and Society' (https:// royalsocietypublishing.org/doi/full/10.1098/rsta.2010.0236)

Brandon Ambrosino, 'If We Made Contact with Aliens, How Would Religions React?' (http://www.bbc.com/future/story/20161215 -if-we-made-contact-with-aliens-how-would-religions-react)

INDEX